AutoCAD 2025 Best Practices, Tips, and Techniques

Boost productivity with interface tips, dynamic blocks, annotations, customizations, and more

Jeanne Aarhus

AutoCAD 2025 Best Practices, Tips, and Techniques

Group Product Manager: Rohit Rajkumar

Publishing Product Manager: Vaideeshwari Muralikrishnan

Book Project Manager: Sonam Pandey

Senior Editor: Rashi Dubey

Technical Editor: Simran Ali

Copy Editor: Safis Editing

Indexer: Tejal Soni

Production Designer: Jyoti Kadam

Senior DevRel Marketing Executive: Nivedita Pandey

First published: October 2024

Production reference: 1100924

Published by Packt Publishing Ltd.

Grosvenor House

11 St Paul's Square

Birmingham

B3 1RB, UK.

ISBN 978-1-83763-672-3

www.packtpub.com

I dedicate this book to my husband, Steve, who has provided me with unwavering support and patience while I spent many hours putting my thoughts to paper over the past year. I have spent the last several years traveling around the world, training and consulting with companies and users, and presenting at conferences while you held down the fort at home. Many Thanks!

– Jeanne Aarhus

Contributors

About the author

Jeanne Aarhus has been involved in computer-aided design for more than 40 years. During this time she has been a designer, trainer, administrator, programmer, and consultant using both Autodesk and Bentley products. She is known for keeping her presentations fast-moving and fun, while providing a thorough understanding of the topic. Her specialty is in providing users with the necessary tools for increasing productivity and getting the job done as efficiently as possible, and she continues to focus on maximizing the user's time and efficiency. Jeanne is a graduate of the University of Nebraska and has been certified in all levels of MicroStation and AutoCAD, which enables her to assist users in moving from one CAD environment to another easily and proficiently. She has worked in both the private and government sectors, and she currently is self-employed and works as a consultant using both Autodesk and Bentley products.

It is easy to overlook the many people who contributed to this writing project, but I would like to thank all of the individuals who contributed ideas, unending support, and technical expertise over the years. Without their contributions, I could not have written this book. I would specifically like to thank the many Autodesk contacts who provided support and answered those endless support questions with professional expertise and a cheerful smile. Many thanks to Jerry Berns and Michael Madrid for their direct help and technical support – I couldn't have done this without your help!

Then there are the folks from Packt Publishing. Without their help, this book would still be a collection of scattered ideas and unorganized thoughts just waiting to be put down on paper.

About the reviewers

Olanrewaju Sulaimon Damilola was born and raised in Lagos, Nigeria. He studied mechanical engineering at Yaba College of Technology and obtained a National Diploma. In 2017, He came across AutoCAD during his internship period and developed a flair for design. He is currently studying the same course at the University of Lagos, Nigeria. With many years of experience, he has used the AutoCAD software to design several 2D and 3D projects that were later brought to life. Also, he has taught and continues to share his CAD knowledge in the use of the software with colleagues and interested learners. He has technically reviewed two other splendid CAD books. He is also very proficient with Autodesk Inventor and Fusion360 and is an expert in SolidWorks.

This book is an all-encompassing material that will give any reader a broad comprehension of the AutoCAD software. It reveals countless commands and processes that will be beneficial to anyone who wants to improve their proficiency in the field. I can proudly say that I am a beneficiary of this collection of CAD guides. A hat-tip to the author for discussing overlooked and less-used features and explaining things in simple terms to facilitate easy assimilation.

Grace Ofunu Ameh is a design engineer with a degree in mechanical engineering, specializing in applied mechanics and design from Federal University of Technology Minna. She has significant experience in manufacturing, project management, and CAD tools such as SolidWorks and AutoCAD. Grace has contributed to various engineering projects and publications, bringing a strong technical background to her work as a reviewer.

Table of Contents

2

Using More of the AutoCAD Interface   41

3

Taking Advantage of Annotation   75

4

Making the Most of Dimensions   115

5

Making Tables Work for You   143

6

Discover More About Blocks   183

7

Discover the New Block Tools   213

8

Learn to Automate with Dynamic Blocks   233

9

Taking Layers to the Next Level   257

10

Enhance Your Knowledge of Reference Files   281

11

Enforcing Your CAD Standards   319

12

Making the Most of Common Commands   333

13

Using Commands to Improve Performance   373

Preface

Learning AutoCAD can be a complex and time-consuming task, and every version brings with it new twists to existing features along with new capabilities. This book contains detailed step-by-step instructions making it possible for any AutoCAD user to get value from this content. Whether you are a new user or an experienced CAD Manager, this book will introduce and clarify many new methods that can be applied to your daily workflows.

Who this book is for

This book was written primarily for the experienced AutoCAD user but can be used by all AutoCAD users as long as you have some experience with the basics. The book includes detailed step-by-step instructions for each tip or technique included. My goal in writing this book was to introduce concepts and commands that are not widely known and often overlooked by many AutoCAD users. During my years of training and consulting on Autodesk products, I have discovered many commands that users tend to overlook or have never discovered, so I wanted to fill this gap using the contents of this book. While no book can be all-inclusive, I predict that many of you will pick up several tips while reading this book. I cannot tell you how many times I have heard the comment "How long has that been there?" or "Why haven't I heard of this before?" when training a group of AutoCAD users. So enjoy the tips and techniques shared in this book, and I will continue to collect new tips and techniques to share in the future!

What this book covers

Chapter 1, Using All of the AutoCAD Interface, lets you learn to use the new interface components of AutoCAD to optimize your working environment.

Chapter 2, Using More of the AutoCAD Interface, lets you learn to use the new and old interface features using different techniques to control the display and functionality of grips, selection sets, and objects.

Chapter 3, Taking Advantage of Annotation, lets you learn to use the full capabilities of the Text Editor and how to create special annotation objects in your drawings.

Chapter 4, Making the Most of Dimensions, lets you learn to improve your efficiency in using the dimension tools using standards and various settings that can help you annotate your drawing more quickly.

Chapter 5, Making Tables Work for You, lets you learn to use the new TABLE objects to collect and generate schedules from the data in your drawing files and from external files using other Windows applications.

Chapter 6, Discover More About Blocks, lets you learn to improve your use of blocks through automation by using lesser-known block commands.

Chapter 7, Discover the New Block Tools, lets you learn how to add the ability to control and automate daily tasks when using both new and old block commands.

Chapter 8, Learn to Automate with Dynamic Blocks, lets you learn to add intelligent and flexible automation to those old blocks using dynamic blocks.

Chapter 9, Taking Layers to the Next Level, lets you learn to manage and use some of the hidden layer commands. Managing your layers is critical to standardizing and simplifying some of the most used commands in AutoCAD.

Chapter 10, Enhance your Knowledge of Reference Files, lets you learn to fully understand what capabilities are available and how they can improve your use of reference files and PDFs.

Chapter 11, Enforcing your CAD Standards, lets you learn to define and enforce your CAD Standards. Standardization can only be achieved through good planning.

Chapter 12, Making the Most of Common Commands, lets you learn how to use some of the most commonly used commands differently. Do you know everything about TRIM, PEDIT, ARRAY, and other commands? This chapter covers these commands in detail.

Chapter 13, Using Commands to Improve Performance, lets you learn how to use the utilities provided in AutoCAD to help you clean up your drawing files and how to repair and audit them when needed.

To get the most out of this book

We assume you are reading this book chronologically. If a topic is explained early in an exercise, it might not be repeated later.

Software/hardware covered in the book	Operating system requirements
AutoCAD 2025 (any other version of AutoCAD will also work)	Windows or macOS

You can refer to the following link to confirm your hardware requirements:

```
https://www.autodesk.com/support/technical/article/caas/sfdcarticles/
sfdcarticles/System-requirements-for-AutoCAD.html?us_oa=dotcom-us&us_
si=9b630dac-e4eb-4e94-b32d-01c1026e2a72&us_st=system%20requirements%20
for%20autocad.
```

I will be using a Windows environment for these examples.

Download the exercise files

You can download the complete profile, resources, and exercises for this book from GitHub at `https://github.com/PacktPublishing/AutoCAD-2025-Best-Practices-Tips-and-Techniques`. If there's an update to the code, it will be updated in the GitHub repository.

We also have other code bundles from our rich catalog of books and videos available at `https://github.com/PacktPublishing/`. Check them out!

Conventions used

There are a number of text conventions used throughout this book.

> **Tips or important notes**
> Appear like this.

This book uses some additional conventions that you should know that will help you understand and follow the steps provided. Here is the formatting found in this document:

- COMMAND NAME: such as LINE, POLYLINE, and ERASE

- *Mouse Controls*: such as *right-click*, *left-click*, and *left-click and drag*

- *Keyboard Shortcuts*: such as *Ctrl + F*, *Shift*, and *Esc*

- `Key-in Commands`: such as `COMMANDS` and `COMMAND OPTIONS`

- Command locations: This table displays the various locations where you can find the command discussed. You should select the command location that best suits your method of using AutoCAD.

    Example:

COMMAND NAME	Command Locations
Ribbon	Ribbon Tab Name | Ribbon Group Name | Command Name
Right-click Menu	
QAT	DDL | COMMAND NAME
Command Line	COMMAND (COMMAND ABBREVIATION)

- System Variables: This table will display each system variable as it is used throughout the book. It includes the description, settings, and all the available variable options.

Example:

VARIABLE NAME	
Description	
Type: Variable Type	
Saved in: Saved Location	
Value 1	Setting description
Value 2	Setting description
Value 3 (default)	Setting description

Get in touch

Feedback from our readers is always welcome.

General feedback: If you have questions about any aspect of this book, email us at `customercare@packtpub.com` and mention the book title in the subject of your message.

Errata: Although we have taken every care to ensure the accuracy of our content, mistakes do happen. If you have found a mistake in this book, we would be grateful if you would report this to us. Please visit `www.packtpub.com/support/errata` and fill in the form.

Piracy: If you come across any illegal copies of our works in any form on the internet, we would be grateful if you would provide us with the location address or website name. Please contact us at `copyright@packtpub.com` with a link to the material.

If you are interested in becoming an author: If there is a topic that you have expertise in and you are interested in either writing or contributing to a book, please visit `authors.packtpub.com`.

Share Your Thoughts

Once you've read *AutoCAD 2025 Best Practices, Tips, and Techniques*, we'd love to hear your thoughts! Scan the QR code below to go straight to the Amazon review page for this book and share your feedback.

https://packt.link/r/1837636729

Your review is important to us and the tech community and will help us make sure we're delivering excellent quality content.

Download a free PDF copy of this book

Thanks for purchasing this book!

Do you like to read on the go but are unable to carry your print books everywhere?

Is your eBook purchase not compatible with the device of your choice?

Don't worry, now with every Packt book you get a DRM-free PDF version of that book at no cost.

Read anywhere, any place, on any device. Search, copy, and paste code from your favorite technical books directly into your application.

The perks don't stop there, you can get exclusive access to discounts, newsletters, and great free content in your inbox daily

Follow these simple steps to get the benefits:

1. Scan the QR code or visit the link below

https://packt.link/free-ebook/9781837636723

2. Submit your proof of purchase
3. That's it! We'll send your free PDF and other benefits to your email directly

1

Using All of the AutoCAD Interface

The out-of-the-box AutoCAD interface is a generic compilation of the most commonly used features, which may or may not be the right combination for you. Whether you are a novice or an experienced user, it is important that you learn the different methods for using this interface.

In this chapter, you will learn how to modify the default interface components and how to expose some hidden interface features to better match your use of the software. You will start by learning how to modify the display and functionality of the default windows. Then, you will see how to optimize your coordinate input to match your use of AutoCAD. Next, you will move on to learn about the usage of hidden shortcut keys. Then, you will see how to control object selections and the associated display options. Finally, you will customize the default Status Bar.

In this chapter, you will cover the following topics:

- Taking advantage of the new features
- Controlling the Command Line
- Useful Dynamic Input
- Using the Clipboard efficiently
- Using Temporary Overrides
- Cycling in AutoCAD
- Hidden in the ViewCube
- Tailoring your Options
- Using File and Layout tabs

By the end of this chapter, you will be able to optimize the interface to work more efficiently within your own work environment.

Technical requirements

You will need a computer with either Windows or macOS to complete this chapter's exercises. I will be using a Windows environment for these examples. Any version of AutoCAD will work, but it is recommended that you use the latest version so your software matches the examples as closely as possible. I will be using AutoCAD 2025 throughout this book, and you can refer to the following link to confirm your hardware requirements: `https://www.autodesk.com/support/technical/article/caas/sfdcarticles/sfdcarticles/System-requirements-for-AutoCAD.html?us_oa=dotcom-us&us_si=9b630dac-e4eb-4e94-b32d-01c1026e2a72&us_st=system%20requirements%20for%20autocad`.

Formatting found in this document

COMMAND NAME: such as LINE, POLYLINE, and ERASE

Mouse Controls: such as *right-click*, *left-click*, and *left-click and drag*

Keyboard Shortcuts: such as *Ctrl + F*, *Shift*, and *Esc*

`Key-in Commands`: such as COMMANDS and COMMAND OPTIONS

Command locations

This table displays the various locations where you can find the command discussed. You should select the command location that best suits your method of using AutoCAD.

Example:

COMMAND NAME	Command Locations
Ribbon	Ribbon Tab Name | Ribbon Group Name | Command Name
Right-click Menu	
QAT	DDL | COMMAND NAME
Command Line	COMMAND (COMMAND ABBREVIATION)

MATCHPROP	Command Locations
Ribbon	Home | Properties | Match Properties
QAT	Drop-Down-List | Match Properties
Command Line	MATCHPROP (MA)

System Variables

This table will display each system variable as it is used throughout the book. It includes the description, settings, and all the available variable options.

Example:

VARIABLE NAME	
Description	
Type: Variable Type	
Saved in: Saved Location	
Value 1	Setting description
Value 2	Setting description
Value 3 (default)	Setting description

SELECTIONPREVIEW	
Controls the display of objects and how they are highlighted when the cursor hovers over them. This highlighting indicates that the object can be selected if you *Left-Click* on it. This setting is stored as a BITCODE that uses the sum of any or all of the following values:	
Type: Bitcode	
Saved in: Registry	
0	OFF (will improve the performance of AutoCAD)
1	ON when no commands are active (the list dialog is not displayed)
2	ON when a command prompts for object selection (the list dialog displays the selected objects that you can cycle through)
3 (default)	ON when you "hover" over an object

To get the most out of this book

We assume you are reading this book chronologically. If a topic is explained early in an exercise, it might not be repeated later.

This book uses some conventions that you should know that will help you understand and follow the steps provided.

You can download the complete profile and exercises on GitHub at the following URL: `https://github.com/PacktPublishing/AutoCAD-2025-Best-Practices-Tips-and-Techniques`.

Using the Exercise Files

The exercise files for this book are located in the ACAD_TipsTechniques\Exercise_Files folder. I recommend placing the ACAD_TipsTechniques folder on your desktop to simplify this installation. You can place them anywhere you want in your system; however, you will need to edit the profile AutoCAD_TipsTechniques.ARG file and import it to use when working with the files for this course.

First, locate the delivered AutoCAD_TipsTechniques.ARG file, a customized profile that directs AutoCAD to use the resources required for this book. This file should be imported to prevent any interference with your working AutoCAD profiles.

1. Download the ACAD_TipsTechniques\Exercise_Files folder for this book and place it on your desktop.

2. Start up the AutoCAD application and open any DWG file.

3. Place the cursor in the middle of the view window and *right-click* to access the **Options** command.

4. Select the **Profiles** tab, and click the **Import** button. Navigate to the delivered profile (.ARG) file located at:

    ```
    ..ACAD_TipsTechniques\Exercise_Files\CompanyXYZ\support\
    ACAD_TipsTechniques.ARG
    ```

5. In the **Import Profile** dialog, leave the name as delivered and select the **Apply** and **Close** buttons.

    This will add the **ACAD_TipsTechniques** profile to the **Available Profiles** list.

6. Select the new profile, **ACAD_TipsTechniques,** and click the **Set Current** button to make it the active profile for working with the files used throughout this book.

> **Note**
>
> Again, using the ACAD_TipsTechniques.arg profile will prevent any interference with your personal or corporate AutoCAD environment.

7. Click **OK** to close the **Options** dialog and save these changes.

Using this profile directs AutoCAD to the correct locations for the various resources used throughout this book.

Taking advantage of the new features

In this section, we will learn how to use the new interface features found in the most recent versions of AutoCAD. Using the newest features can improve your overall use of the software.

First, let's examine the new "floating" or "undocked" drawing windows to take advantage of large and multiple monitor configurations.

Floating Drawing Windows

In this exercise, we will discuss how to take advantage of using more than one monitor. With multiple monitors, you might prefer to "float" your drawing windows to take advantage of that second monitor. The benefit of a "floating" drawing window may not be immediately apparent, so let me demonstrate how it can help you with your workflow:

- You can take advantage of multiple monitors by placing separate drawing windows on each monitor. This is more efficient than just stretching a single drawing window across multiple monitors.

- You can use floating windows to run commands between two drawing files.

Each of the floating windows has its own Command Line dialog. To demonstrate this feature, we need to have two drawings open and follow these steps:

1. Open the 1-1_Floating Windows 1.dwg and 1-1_Floating Windows 2.dwg files. Both files are now visible in the **File Tabs** area.

2. Using the **File Tabs** area, *left-click and drag* the 1-1_Floating Windows 1.dwg File tab into the view window and release the mouse button.

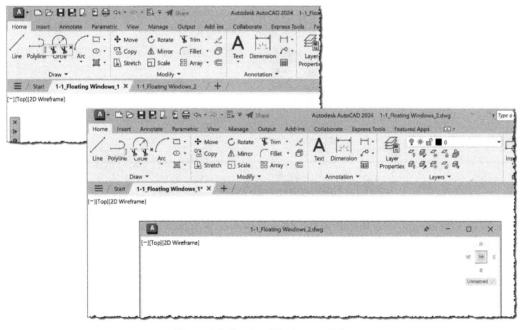

Figure 1.1: Floating Windows or Tabs

3. This file is now a "floating" drawing window that can be moved to either monitor and maximized or tiled as needed.

> **Note**
>
> This is a great feature for those familiar with other applications that take advantage of multiple monitors.

In the next exercise, we will examine how to PIN and RESTORE the "floating" view windows.

Pin and Restore

Once the file window is "floating," you can control the display order of the window using one of the following methods:

- Pin the windows to control the display order
- Tile the windows side by side

In this exercise, we will discover how to PIN and TILE the "floating" drawing view windows:

1. First, use the **PIN** icon 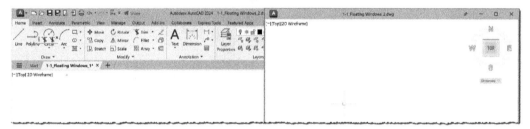 found in the floating file's Title Bar to force that file on TOP of the AutoCAD application. The **PIN** icon will change to a blue-colored icon when activated. Resize the "floating" file window to one-half of the screen display. This prevents you from using the *Ctrl + Tab* shortcut key to toggle between open drawing files.

Figure 1.2: Floating drawing window in TILE mode

2. Next, *left-click and drag* the file to the RIGHT side of the screen until you see it lock in place in TILE mode.

3. *Left-click and drag* the `1-1_Floating Windows 2.dwg` file to the LEFT side of the screen until you see it lock in place in TILE mode.

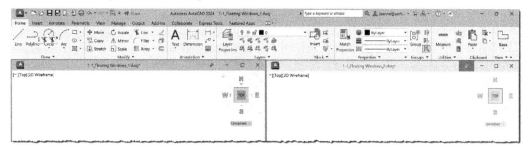

Figure 1.3: Floating drawing windows resized for ribbon display

4. Resize both files to see the AutoCAD application ribbon bar.

5. Use the **PIN** icon on both files to force the drawing view windows to remain above the application window.

MATCHPROP	Command Locations
Ribbon	Home | Properties | Match Properties
QAT	Drop-Down List | Match Properties
Command Line	MATCHPROP (MA)

6. Using the **Quick Access Toolbar (QAT)**, turn ON the icon for the **Match Properties** command.

7. Use the **QAT** *drop-down list* to add a checkmark to the **Match Properties** command.

Figure 1.4: Quick Access Toolbar | Match Properties

8. Select the `1-1_Floating Windows 1.dwg` file and select the **Match Properties** command from the **QAT**. *Left-click* on a ROOM NAME (**LAUNDRY**) object to match the properties of the text object in this file.

9. Select the `1-1 Floating Windows 2.dwg` file and select a ROOM NAME (**BEDROOM**) text object in this view. The text object in this file will change to match the text object properties from the first file.

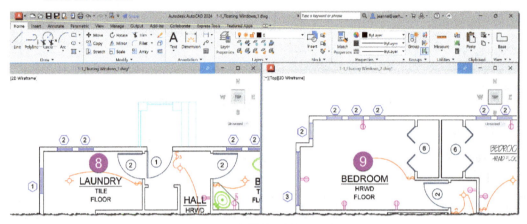

Figure 1.5: QAT Match Properties

You can re-dock a "floating" drawing window using *left-click and drag* to drop the file's Title Bar back into the **File Tabs** area and then release the mouse button.

> **Note**
>
> Did you notice that each file has its own Command Line when working with "floating" view windows?

You can control this new feature using the SYSFLOATING system variable:

SYSFLOATING	
Controls the dock state of the drawing File Tabs.	
Type: Integer	
Saved in: Registry	
0 (default)	Drawing File Tabs are docked.
1	Drawing File Tabs are floating.
-1	The floating drawing File Tab feature is disabled.

On completion of this exercise, you should now be able to control the display of the drawing to take advantage of your hardware configuration and customize your QAT to include the MATCH PROPERTIES command.

Controlling the Command Line

In this exercise, we will learn how to control where your Command Line dialog "lives" and how to find it when it goes missing.

Lost Command Line?

Have you ever lost the Command Line window? I guess you have, since it occasionally disappears while working in the interface. You can use the following two options to recover the Command Line from its hiding place.

Option 1 – Using a Keyboard Shortcut

The first option is to use the built-in *Ctrl + 9* shortcut key to toggle the command window ON and OFF. This will bring the Command Line dialog to the front of the application window from wherever it is hiding.

Option 2 – Using the CUI Dialog

The second option is to use the Customize User Interface (CUI) command to turn the command window ON or OFF.

The CUI command is available at the following locations:

CUI	Command Locations
Ribbon	Manage | Customization | User Interface
Command Line	CUI (CUI)

Using the following steps, you can gain control of your Command Line dialog. To demonstrate this functionality, you need to be in a drawing file:

1. Open the 1-0_Blank.DWG file.
2. Start the CUI command and, using the **Customize User Interface** dialog, select the **(current)** workspace.
3. Click the **Customize Workspace** button at the TOP of the RIGHT panel. This will turn all the menu items **blue**.
4. Expand the **Palettes** item and select the **Command Line** palette.

5. Using the **Properties** panel in this dialog, modify the **Orientation** setting to a different orientation using the **Floating**, **Top**, **Bottom**, **Left**, **Right**, or **Do Not Change** options. Click the **Apply** button and close the dialog.

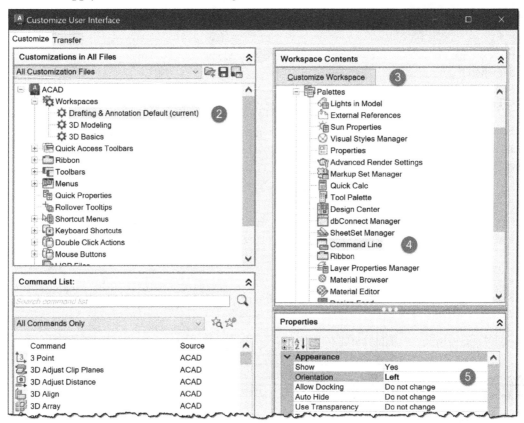

Figure 1.6: CUI dialog results

The Command Line should appear at your selected location. I selected the LEFT command-line location.

On completing this exercise, you can control the location and display properties of your Command Line to improve your overall use of your screen real estate.

Does your Command Line keep moving?

Are you having problems with your Command Line not "sticking" at your preferred location? Me too, and it can be very annoying!

Locate and size your Command Line as needed and use the *Ctrl* key when dragging it to the desired location. Holding the *Ctrl* key while moving the Command Line dialog disables the "magnet" or "docking" capability for the positioning of the Command Line dialog and forces the location to be memorized.

Use the following steps to prevent the Command Line dialog from moving around in the interface:

1. Continue using the 1-0_Blank.dwg file.

2. Select the Command Line using the "dotted" end of the title bar, then *left-click and drag* it to a new location while holding the *Ctrl* key. If you *drag* the dialog down into the Status Bar area of the interface, you will see that it no longer forces the "docking" functionality.

3. *Drag* the **Command Line** dialog to your preferred location and re-size it to one line to minimize the required screen real estate. As an experienced AutoCAD user, you probably don't need to see more than one line of the Command Line dialog, and you can always use the *F2* key to review additional lines when needed.

Control the Command Line Search Options

You can control what commands appear in the Command Line suggestion list using the INPUTSEARCHOPTIONS system variable. This variable allows you to control the following options.

AutoComplete

Controls if the commands are automatically completed as you type. You can determine if a MID-STRING KEY-IN is allowed or if you wish to have the suggestion list sorted based on the FREQUENCY OF COMMAND USAGE or ALPHABETICALLY.

AutoCorrect

This enables spellchecking on the commands as you type.

System Variables

This controls whether system variables are included in the suggestion list. When they are, you need to press *Tab* key to expand the list.

List Time Delay

This controls how long the delay is before the suggestion list is displayed as you type. This time is entered in milliseconds.

Filter Named Objects

This controls whether named objects, such as blocks, layers, hatches, and styles, are displayed in the suggestion list. You can determine which types are included and control their priority by using the arrow buttons to change the order of the list.

In this exercise, we will learn how to customize the Command Line location and features.

1. Continue using the 1-0_Blank.dwg file.

2. Using the Command Line, *left-click* on the WRENCH icon 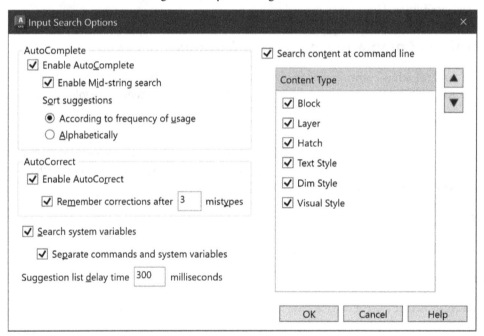 to access the **Input Search Options...** command.

3. Using the **Insert Search Options** dialog, review the settings and make any desired changes.

4. Click **OK** to close the dialog and save your changes.

Figure 1.7: INPUTSEARCHOPTIONS settings

After completing this exercise, you will be able to control the location of your Command Line with more precision and disable the "docking" functionality when needed to make the most of your screen real estate. Using these skills, you can customize what content is displayed in your Command Line using the suggestion list options.

Useful Dynamic Input

In this exercise, we will learn how to use Dynamic Input and clipboard shortcuts to improve your productivity using precision input and Dynamic Input. First, you need to know your dynamic input rules.

Know your Dynamic Input rules

Have you ever wanted to input Absolute coordinates using the new Dynamic Input feature? First, let's verify that you have Dynamic Input enabled:

1. Open the 1-2_Know Your Coordinates.dwg file.
2. Using the Status Bar, *left-click* on the "hamburger" icon, ☰, and select **Dynamic Input** from the list to turn it ON in the Status Bar.
3. This will turn ON the **Dynamic input** icon, ⊞.

First, let's review the basics of AutoCAD coordinate input:

- **Relative Coordinates**: 2,20 is relative to the previous coordinate location
- **Absolute Coordinates**: #2,20 is an absolute coordinate location in the current UCS (User Coordinate System),
- **World Coordinates**: *2,20 is a world coordinate location

Once you understand these rules, you can input them using two methods.

Method 1

For the first example, we want to add an object vertex at the absolute coordinate of 0,0 using Dynamic Input. When using Dynamic Input, if you key in 0,0 the coordinate is interpreted as a "relative" coordinate from the current cursor location. If you key in #0,0 it is interpreted as an "absolute" coordinate.

Use this override setting when inputting the coordinate values by typing in a # character before the coordinate:

1. Select the **Line** command and *left-click* inside the SQUARE object to start the line.
2. *Drag* the cursor to the LEFT side of the SQUARE object, key in #0,0 and press *Enter* to complete the command. Notice the LINE is not continued using this key in is interpreted as the ABSOLUTE coordinate 0,0.
3. *Drag* the cursor to the TOP edge of the SQUARE object, key in 0,0 and press *Enter* to complete the command. Notice that the line is not continued using this key-in, as it is interpreted as X=0 and Y=0 from the current location.
4. Press the *Esc* to cancel the current command and *Ctrl + Z* shortcut key to UNDO the previous lines.

Next, let's look at another method to input coordinates.

Method 2

If you key in ABSOLUTE coordinates more than RELATIVE coordinates, you will want to change your input settings when using Dynamic Input to avoid using all the # characters during input.

Let's change our default input settings before we draw the next line. Follow these steps:

1. Continue using the `1-2_Know Your Coordinates.dwg` file.

2. To access the Dynamic Input Settings, *right-click* on the Dynamic Input icon 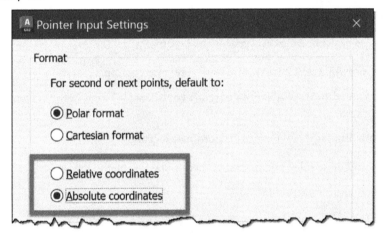 in the Status Bar and select the **Dynamic Input Settings** command.

3. Using the Enable Pointer Input, select the **Settings** button.

4. Using the **Pointer Input Settings** dialog, select the **Absolute coordinates** format for our X,Y,Z key-ins.

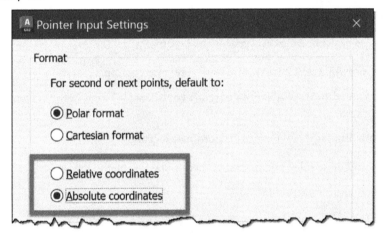

Figure 1.8: Absolute Pointer Input Settings

5. Click **OK** to save your selection and click **OK** again to close the **Dynamic Input Settings** dialog.

6. Select the LINE command and *left-click* inside the SQUARE object to start the line.

7. *Drag* the cursor to the LEFT side of the SQUARE, key in 0 , 0 and press *Enter* to complete the command. This time, the LINE is drawn to the ABSOLUTE X,Y coordinate without using the # prefix.

After completing this section, you can use these rules and settings for your coordinate input. You can control and simplify your coordinates more easily and still take advantage of Dynamic Input.

Using the Clipboard efficiently

Everyone knows you can use the *Ctrl + C* Windows shortcut keys to copy objects to the clipboard and *Ctrl + V* to paste them back into a drawing. These are wonderful shortcuts, but they have a "flaw" in our AutoCAD workflow. These methods don't allow you to define a base point to control the insertion point for the paste portion of these actions. Instead of using the typical Windows shortcuts, use the COPY WITH BASE POINT clipboard command to control your base point.

CLIPBOARD	Command Locations
Ribbon	Home | Clipboard | Copy | Copy with Base Point
Command Line	COPYBASE (COPYB)
Right-Click Menu	Clipboard | Copy with Base Point

Copy with a base point

In this exercise, we will use the *Ctrl + Shift + C* shortcut to define a base point and copy objects to the clipboard, which is quicker than using the *right-click* menu to access this command:

1. Open the `1-3_Smarter Clipboard.dwg` file.
2. Select the ARROW object and *right-click* to access the **Clipboard | Copy With Base Point** command, or use the *Ctrl + Shift + C*.
3. Using the ENDPOINT OSNAP, *left-click* at the tip of the arrowhead to define the base point.
4. Next, use the *Ctrl + V* shortcut to PASTE the clipboard contents into the drawing using the newly defined base point.

> **Note**
>
> You can also use the *Ctrl + Shift + X* shortcut to CUT objects from a drawing to the clipboard and define a base point.

Paste and rotate from the Clipboard

Another clipboard option that is "hidden" from many users is the ability to rotate the clipboard contents during PASTE operation:

1. Continue using the 1-3_Smarter Clipboard.dwg file.

2. Use the *Ctrl + V* shortcut to PASTE the ARROW graphics again, and key in R to pre-define the desired angle of the ARROW graphics.

> **Note**
>
> The "R" command option is "hidden" and is NOT displayed in the Command Line but is available to use.

3. Key in 90 to rotate the new ARROW graphics 90 degrees and *left-click* in the drawing view to place the new rotated graphics.

Figure 1.9: Paste and rotate

Paste as a block from the Clipboard

In this exercise, we will continue to use PASTE from the clipboard contents and convert all graphics as a BLOCK object during the PASTE process:

1. Continue using the 1-3_Smarter Clipboard.dwg file.

2. Use the *Ctrl + Shift + V* shortcut to PASTE the clipboard graphics as a BLOCK.

3. The resulting BLOCK object will be an ANONYMOUS BLOCK with an **A$C....** block name.

4. Using the Command Line, key in RENAME to give this new block a logical name.

RENAME	Command Locations
Command Line	RENAME (REN)

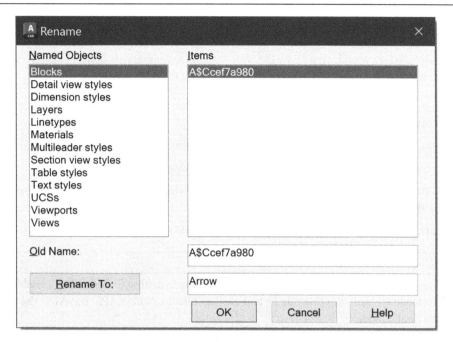

Figure 1.10: Renaming the block

Upon completing this section, you should be able to use the clipboard more efficiently to copy and paste drawing objects and place them correctly during the clipboard commands.

Using Temporary Overrides

In this section, we will use several keyboard overrides to override command functionality on the fly to help control the placement and editing of objects in the drawing file.

Shift for ORTHO

In this first exercise, we will learn to use the *Shift* key to automatically apply the ORTHOGONAL (ORTHO) angle lock without using the *F8* function key. The *Shift* key will apply ORTHO as a temporary override while executing a command:

1. Open the 1-4_Shortcuts.dwg file.

2. Select the **Line** command and *left-click* in the view window to start the LINE object.

3. *Drag* the cursor to the RIGHT and UP at approximately 45 degrees. By default, POLAR is turned ON, and you can draw the line at any angle.

4. Hold down the *Shift* key to force ORTHO to be applied, limiting your angle options to "0" or "90" degree angles.

5. Release the *Shift* key to return to the previous POLAR angle options.

Shift + A for OSNAPS

In this second exercise, we will learn to use the *Shift + A* keys to temporarily disable OSNAPS without using the *F9* function key. The *Shift + A* key will disable the OSNAPS as a temporary override while executing a command:

1.   Continue using the 1-4_Shortcuts.dwg file.

2.   Select the **Line** command and *left-click* in the view window to start the LINE object.

3.   *Drag* the cursor to the UPPER-LEFT corner of the shape.

4.   Hold down the *Shift + A* keys to disable OSNAP, allowing you to avoid accidentally snapping to the shape object.

5.   Release the *Shift + A* keys to enable OSNAPS again when needed.

OSMODE	
Defines what OSNAPS are set in running mode. For example, setting OSMODE to 7 enables the Endpoint (bitcode 1), Midpoint (bitcode 2), and Center (bitcode 4) OSNAPS. You ADD the values to define the final bitcode.	
Type: Bitmode	
Saved in: Registry	
Initial Value: 4133 (Endpoint, Center, Intersection, and Extension)	
0	None
1	Endpoint
2	Midpoint
4	Center
8	Node
16	Quadrant
32	Intersection
64	Insertion
128	Perpendicular
256	Tangent
512	Nearest
1024	Geometric Center
2048	Apparent Intersection
4096	Extension
8192	Parallel
16384	Suppresses the current running OSNAPS

Cycling in AutoCAD

For several years now, the ability to perform object cycling between objects using the mouse and keyboard shortcuts has been available to assist in selecting specific overlapping objects. In the newer versions of AutoCAD, you can use the new SELECTION CYCLING tool available in the Status Bar. In the next example, let's look at both methods for object cycling.

Object Cycling "the Legacy way"

When you *hover* over overlapping objects, you can use the *Shift + Spacebar* shortcut to cycle between all the overlapping objects. Once the object you need is highlighted, you can *left-click* to select the Selection Cycling command.

By default, the SELECTIONPREVIEW system variable is set to 3, which allows a preview of the object under the cursor location. Other settings are available, as shown in the following table:

SELECTIONPREVIEW	
Controls the display of objects and how they are highlighted when the cursor hovers over them. This highlighting indicates that the object can be selected if you *Left-Click* on it. This setting is stored as a BITCODE that uses the sum of any or all of the following values:	
Type: Bitcode	
Saved in: Registry	
0	OFF (will improve the performance of AutoCAD)
1	ON when no commands are active (the list dialog does not display)
2	ON when a command prompts for object selection (the list dialog displays the selected objects that you can cycle through)
3 (default)	ON when you "hover" over an object

Use the following steps to use the "legacy" method for cycling between overlapping objects:

1. Open the `1-5_Object Cycling.dwg` file.

2. *Hover* the mouse over the overlapping objects at **P1** and use *Shift + Spacebar* to cycle between the objects at that location. You will find that there are four objects at this location:

    LINE, LINE, POLYLINE, POLYLINE

3. Continue to use the *Shift + Spacebar* to toggle the highlight between all four objects. When the object you want to select is highlighted, release *Shift + Spacebar* and *left-click* to select that object.

4. Press the *Esc* key to clear the active selection set.

Next, let's learn how to use the new method for controlling the selection of overlapping objects.

Object Cycling "the New way"

In this exercise, we will look at the dialog option provided by newer versions of AutoCAD to assist with selecting overlapping objects using the SELECTION CYCLING tool in the Status Bar. Use this method if you struggle to see the highlighted objects demonstrated in the previous method:

1. Continue using the 1-5_Object Cycling.dwg file.

2. By default, the **Selection Cycling** tool is turned OFF. Using the Status Bar, *left-click* on the "hamburger" icon ☰ to turn on **Selection Cycling** in the Status Bar. †

3. *Hover* the mouse over the overlapping objects at **P1** and make note of the **Cursor Badge** ⬚ that informs you of the overlapping objects at this point.

4. Select the overlapping objects at **P1**, and a **Selection** dialog opens to display all objects at that location.

> **Note**
>
> You can turn OFF the display of the cursor badges using the CURSORBADGE system variable.

CURSORBADGE	
Determines which cursor badges are displayed in the drawing area when hovering on an object.	
Type: Integer	
Saved in: Registry	
1	Turns OFF cursor badges used in the following commands: AREA, COPY, DIST, ERASE, ID, LIST, MASSPROP, MEASUREGEOM, MOVE, ROTATE, SCALE, TRIM, and ZOOM
2 (default)	Turns ON all cursor badges

Selection Cycling Settings

You can control how the selection cycling dialog appears and what it contains using the Selection Cycling Settings:

1. Using the Status Bar, *right-click* on the **Selection Cycling** button † to access the **Selection Cycling Settings**.

2. Using the **Drafting Settings** dialog, use the **Quadrant** and **Distance** settings to define where and how far away the dialog appears from your cursor. You can also choose the **Static** location, and the dialog will appear at the same location regardless of your cursor location.

3. You may also prefer to turn OFF the **Title Bar**, as it is unnecessary.

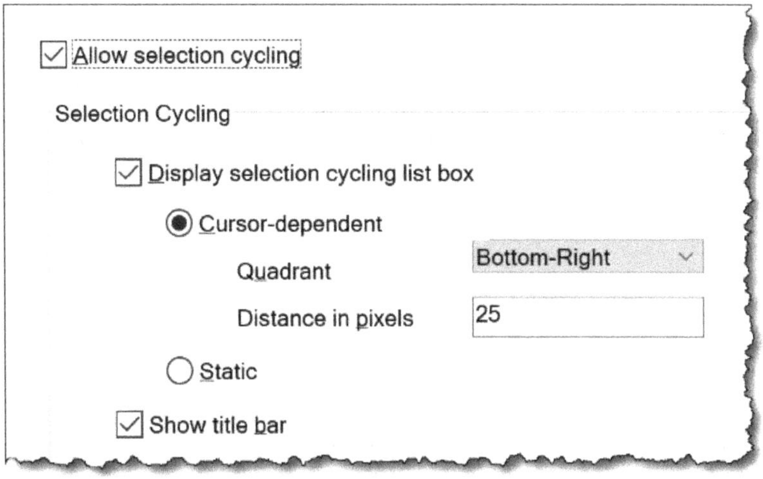

Figure 1.11: Selection Cycling settings

In this exercise, we learned how to better control the selection of overlapping objects using both legacy and new methods.

Next, let's learn how to control OSNAP Cycling.

OSNAP Cycling

In this exercise, let's investigate how to control the use of our running OSNAPS to better control which OSNAP method is applied. Do you know all the shortcuts available to simplify your daily use of OSNAPS?

Before we begin these next few examples, we need to verify which OSNAPs are currently running.

1. Open the `1-6_OSNAP Cycling.dwg` file.

2. Using the Status Bar, *left-click* on the OSNAP icon ⬚ ▾ *drop-down-list* and verify that you have the following OSNAPS turned ON:

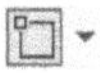

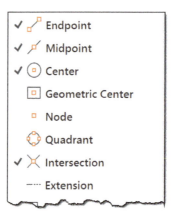

Figure 1.12 Running OSNAPS

3. For this example, we must turn OFF the DYNAMIC INPUT setting $^{+}$ in the Status Bar. The DYNAMIC INPUT functionality prevents this method from working.

4. Select the **Line** command and *hover* the mouse over the **Center** point of the circle.

5. Use the *Tab* key to toggle between the various running OSNAPS. Each object and OSNAP mode will display and highlight to confirm which snappable point is identified. Issue a *left-click* when the correct snappable point and object are displayed.

In this exercise, you learned how to control your OBJECT and OSNAP Cycling to gain more control over your drawing and editing processes.

File Cycling

In this exercise, let's look at the options for switching between open files in an AutoCAD session. If screen real estate is a problem, you can turn OFF the File Tabs and use the *Ctrl + Tab* shortcut to cycle between open files.

First, how do you turn OFF File Tabs? Use the following FILETABCLOSE and FILETAB commands to control the display of File Tabs.

FILETABCLOSE, FILETAB	Command Locations
Ribbon	View	Interface	File Tabs
Command Line	FILETABCLOSE, FILETAB

1. Open the 1-7_File Cycling 1.dwg and 1-7_File Cycling 2.dwg files.

You can use the *Ctrl + Tab*, *Ctrl + Shift + Tab*, and *Ctrl + Home* shortcuts to navigate through the File Tabs. Let's look at these commands in detail:

- *Ctrl + Tab*: Use these shortcut keys to cycle to the next open File Tab.

- *Ctrl + Shift + Tab*: Use these shortcut keys to cycle to the previous open File Tab.

> **Note**
>
> Use the FILETAB command to turn the display of the File Tabs on again.

Using these keyboard shortcuts will simplify your navigation of open files and eliminate the need to display File Tabs, letting you save that screen real estate.

Viewport Cycling

Have you ever been stuck in a viewport that was erroneously or intentionally created inside of another viewport, creating a "nested" viewport? How do you access a "nested" viewport?

1. Open the `1-8_Viewport Cycling.dwg` file.

2. *Double left-click* in the larger viewport to activate that viewport.

    If you try to *double-left-click* to activate the smaller viewport, it doesn't work. However, if you use the *Ctrl + R* shortcut, you can toggle between all viewports, including the embedded viewport, to gain the ability to activate any viewport in this paper space.

3. Use the *Ctrl + R* shortcut to activate the smaller viewport that is embedded in the larger viewport.

4. Continue using *Ctrl + R* to toggle between viewports in this layout.

In this exercise, you learned how to control your FILE and VIEWPORT cycling to gain more control over your drawing and editing process.

Hidden in the ViewCube

Not everyone uses the ViewCube in AutoCAD, but even if you do, there are some hidden features that you may not be aware of. Most users use the ViewCube for 3D work, but did you know you can also use it for 2D?

Let me demonstrate how you can take advantage of the ViewCube's hidden features, even in 2D.

Zoom to Selected Objects

Do you know how to get AutoCAD to perform a ZOOM EXTENTS to a selection only? Using the ViewCube, you can ZOOM EXTENTS to just the selected objects.

1. Open the 1-9_ViewCube 2D.dwg file.

2. Using the **Lasso** selection method, select **Lots 9-15** between **RAY BLVD** and **HAYDEN DR**.

> **Note**
>
> If you are not familiar with the LASSO selection option, *left-click and drag* the cursor around the objects to select. Release the *left-click* when the bounding selection box is complete.

3. *Double left-click* to access the **Zoom Extents** command. As anticipated, it disregards the selected items and zooms out to show the entire drawing's contents.

4. Repeat the previous selection command, and using the ViewCube, select the TOP face. This time, the selected objects are recognized and the view zooms to just the selected objects.

> **Note**
>
> Disable the ZOOM EXTENTS functionality using the ViewCube settings in the next section.

Save View as Home

In this exercise, we will learn how to save a common view of a drawing so that it can be easily recalled. Yes, we could use Saved Views, but there is also another method for saving common views.

Let's begin by discovering how to use the ViewCube to create a Home view quickly.

1. Continue using the 1-9_ViewCube 2D.dwg file.

2. Using the mouse, *double-click* the wheel to access the **Zoom Extents** command which will fit the drawing contents in the current view.

3. Using the **Zoom Window** command, zoom in on **Lots 8-13** between **HAYDEN DR** and **STRATTON DR**.

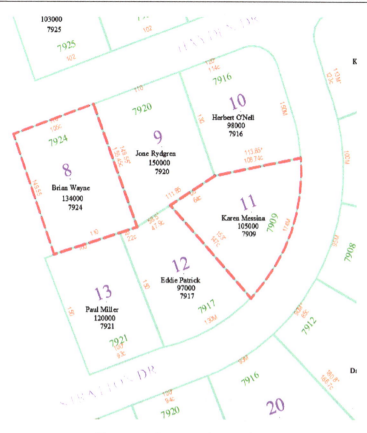

Figure 1.13: Zoom window results

4. *Hover* the mouse over the ViewCube until you see the **Home** icon, 🏠, then *left-click* to select it. By default, it will rotate to an isometric view, which would be great if we were working in a 3D file.

5. Use the **Undo** command (*Ctrl + Z*) to return the view to the previous flat view.

6. Again, *hover* over the ViewCube and *right-click* on the **Home** icon. Select the **Set Current View As Home** command.

7. *Double-click* the mouse wheel to access the **Zoom Extents** command.

8. Use the new **Home** icon definition to set the current view to your preferred Home view.

Now, we can look at the remaining ViewCube customizable settings to suit your preferences.

ViewCube Settings

Use the **ViewCube Settings** to control the display and functionality of the ViewCube:

1. Continue using the `1-9_ViewCube 2D.dwg` file.

2. *Hover* the mouse over the ViewCube and *right-click* to access the **ViewCube Settings** command.

Figure 1.14: Default ViewCube

3. Using the **ViewCube Settings** dialog, use the following settings to change the functionality of the ViewCube:

4. Use the **On-screen position** to control where the ViewCube is displayed. The TOP RIGHT location may not be best for your use.

5. Use the **ViewCube Size** settings to change the overall size of the ViewCube in your view window.

6. Use the **Show UCS Menu** toggle to turn OFF the UCS *drop-down list* if you do not use the UCS.

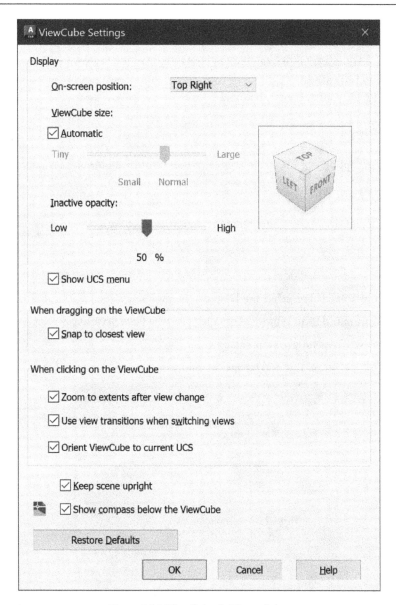

Figure 1.15: ViewCube Settings dialog

In this section, you learned how to take advantage of the hidden features and settings in the ViewCube.

In the next section, we will look at how you can change the appearance of AutoCAD using the OPTIONS command.

Tailoring your Options

Do you need to use AutoCAD for presentations at work? The OPTIONS settings have several controls that you can use to improve the visibility of your cursor and command prompts, making those presentations easier for your audience to follow.

Colors and Sizes

First, let's look at the options for the size and color of the Command Line and Dynamic Prompts.

Command Line Font and Size

To modify the font and size of the Command Line, follow these steps:

1. Open the `1-10_Options.dwg` file.
2. Place the cursor in the middle of the view window and *right-click* to access the **Options** command.
3. Using the **Options** dialog, select the **Display** tab, then select the **Fonts...** button.
4. Using the **Command Line Window Font** dialog, set the **Font Style** and **Size** options as needed.
5. Click **Apply & Close** to save your changes.
6. Click **OK** to close the dialog and review the changes to your Command Line.

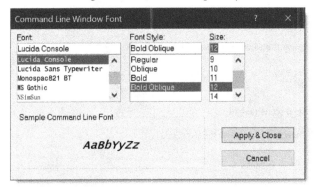

Figure 1.16: Command line window font

Command Line Color

To modify the color of the Command Line, follow these steps:

1. Continue using the 1-10_Options.dwg file.

2. Using the **Options** dialog, select the **Display** tab, then click the **Colors...** button.

3. Using the **Context:** section, select **Command Line**, and using the **Interface Element** section, select the **Active prompt background** (WHITE by default). Change this background color to stand out as needed. I will change it to a YELLOW highlight color.

4. Using the **Color:** *drop-down list*, select the **Select Color...** option, and choose your preferred color. Select a base color and use the **True Color** tab to modify the actual color using R,G,B values. I used 255,255,185 for this example.

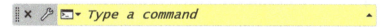

Figure 1.17: Command line highlight color

5. Click **OK** to close the dialog.

6. Click **Apply & Close** and **OK** to save your changes.

Next, let's modify the size and color of your Dynamic Input appearance.

Dynamic Input size and color

I recommend changing the color of your Dynamic Input as well as the Command Line:

1. Continue using the 1-10_Options.dwg file.

2. Using the Status Bar, *right-click* on the **Dynamic Input** icon and select the **Dynamic Input Settings** command.

3. Using the **Drafting Settings** dialog, select the **Drafting Tooltip Appearance** button.

4. Select the **Colors...** button, and using the **Drawing Windows Color** dialog, select the **Context | 2D model space** and **Interface element | Drafting tooltip background**.

5. Using the **Color:** *drop-down list*, select your preferred color using the **Select Color...** option. Select the base color and use the **True Color** tab to tweak the actual color using R,G,B values. I used 250,210,250 for this example.

6. Click **Apply & Close** to save your changes.

7. Use the **Size** setting to increase the text size for the Dynamic Input prompts.

8. Modify these settings to apply to the **Dynamic Input tooltips** only.

9. Click **OK** to save these changes and close the dialog, then **OK** again to complete the changes.

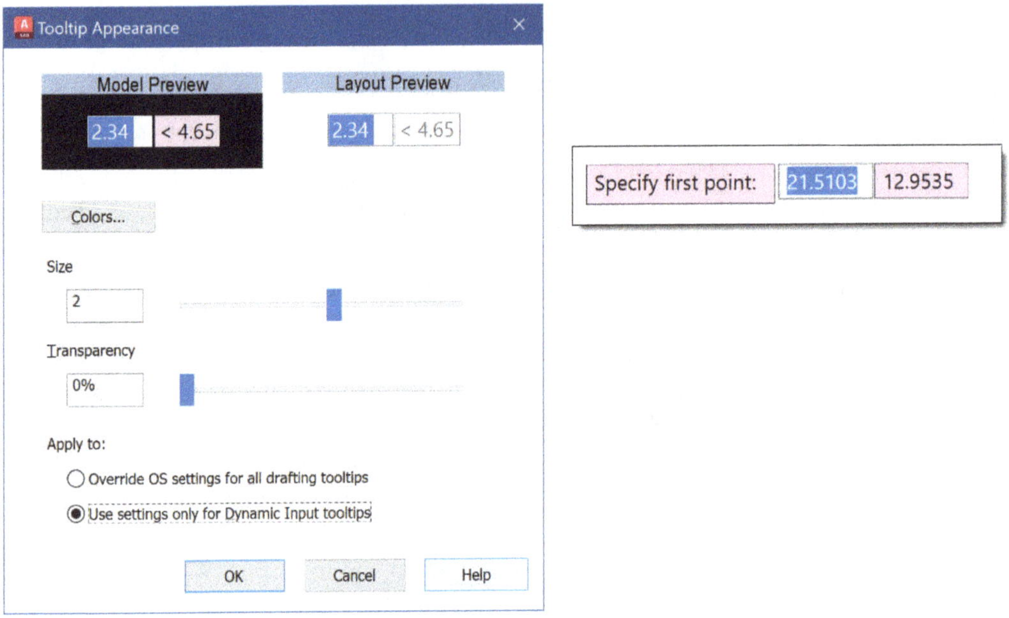

Figure 1.18: Dynamic Input color and size

Next, we will modify the color of the polar tracking and auto-tracking lines so they are easier to see with a dark background.

Polar tracking and auto-tracking lines color

In this exercise, we will modify the color of the polar tracking and auto-tracking lines as needed to make them more visible against your background:

1. Continue using the `1-10_Options.dwg` file.

2. Using the **Options** dialog, select the **Display tab** and select the **Colors...** button.

3. Using the **Context** section, select **2D Model Space**. Then, using the **Interface Element** section, select **Autotrack vector**. Change the background color to one that stands out against your background view color (1,152,1 is the default). I will change it to a yellow highlight color that stands out nicely against the default dark gray background.

4. Using the **Colors...** *drop-down list*, select the **Select Color...** option, and choose your preferred color. Select the base color and use the **True Color** tab to tweak the actual color using R,G,B values. I used `255,255,185` for this example.

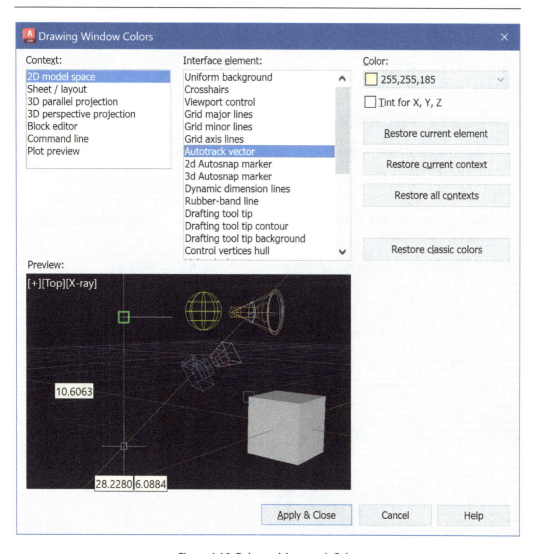

Figure 1.19: Polar and Autotrack Color

> **Note**
>
> You should also consider modifying the 2D and 3D Autosnap markers. It's worth exploring all the options available.

In this exercise, you learned to modify your AutoCAD display options to improve your presentations and possibly your daily work environment by changing the size and color of many aspects of the interface.

Using File Tabs

In this exercise, you will learn to use all the features of the File Tabs to reduce the number of clicks required and to save time.

Close All and Close All Other Drawings

When working on a project, it's common to find you have more files open than you intended. Here's how to solve that problem quickly:

1. Open the 1-11_Using Tabs 1.dwg and 1-11_Using Tabs 2.dwg files.
2. *Right-click* on any drawing **File Tab**, and select **Close All** or **Close All Other Drawings** to quickly close the files you are finished working with.

Another command located on the File Tab that is easily overlooked is the SAVE ALL command.

Save All

Use the SAVE ALL command to quickly save all open drawings:

1. Open the 1-11_Using Tabs 1.dwg and 1-11_Using Tabs 2.dwg files.
2. *Right-click* on any drawing **File Tab**, and select **Save All** to save all open files.

Figure 1.20: File Tab commands

Another command located on the File Tab is the OPEN FILE LOCATION command.

Open File Location

Use the OPEN FILE LOCATION command in the OPEN WINDOWS EXPLORER command. This will open a new Windows Explorer dialog in the project file location:

1. Continue using the 1-11_Using Tabs 1.dwg and 1-11_Using Tabs 2.dwg files.
2. *Right-click* on any drawing File Tab and select **Open File Location** to open a Windows Explorer dialog in that file's location.

Another command on the File Tab is the COPY FULL PATH command.

Copy Full Path

Use the COPY FULL PATH command to copy the drawing file path to the clipboard, which can then be used in documentation or other Windows dialogs:

1. Continue using the 1-11_Using Tabs 1.dwg and 1-11_Using Tabs 2.dwg files.
2. *Right-click* on any drawing File Tab, and select **Copy Full Path** to capture the drawing file location to the clipboard.

You can also use the new File Tab menu in AutoCAD 2024 to access the various file commands.

File Tab menu

The new File Tab menu can be used to switch between open drawings and create, open, save all, or close drawing files:

1. Continue using the 1-11_Using Tabs 1.dwg and 1-11_Using Tabs 2.dwg files.
2. *Left-click* on the "hamburger" icon ≡ tab to access the new File Tab menu.

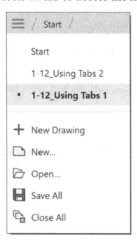

Figure 1.21: New File Tabs menu

By using these file commands, you get access to updated and more efficient options for managing frequently used file commands.

In the next section, we will learn to fully use Layout Tabs.

Using Layout Tabs

In this section, we will discover new options for controlling the appearance of the Layout Tabs. Are you using all the features of your Layout Tabs? Using these features can save you considerable clicks.

Dock above Status Bar

Use the DOCK ABOVE STATUS BAR option to separate the Layout Tabs from the Status Bar. If your project files contain multiple layouts and you need more space to display them all, separate the Layout Tabs from the Status Bar.

Dock Inline with Status Bar

To save screen space and have both the Layout Tabs and Status Bar share the same screen space, use DOCK INLINE WITH STATUS BAR:

1. Open the 1-12_Using Layout Tabs.dwg file.

2. *Right-click* on any Layout Tab, and select **Dock above Status Bar**.

Figure 1.22: Default Layouts docked above the Status Bar

Here is how the modified layouts appear:

Figure 1.23: Modified Layouts docked inline with the Status Bar

> **Note:**
> When using the Dock Inline with Status Bar, you may need to use the Layouts "hamburger" ≡ button to swap between layouts.

Quickly Copy or create New Layouts

You can quickly COPY or create NEW Layout Tabs using the *Ctrl* key:

1. Continue using the `1-12_Using Layout Tabs.dwg` file.

    For this example, we want to change the layout to Dock Above Status Bar to display the results more visually.

2. *Right-click* on the **Architectural** Layout and select the **Dock Above Status Bar** command.

3. Select the **Architectural** tab and hold down *Ctrl + left-click* to select the **Architectural** Layout Tab. *Left-click and drag* the layout copy to the desired location and release *Ctrl + left-click*.

4. You can *double-left-click* on the new layout to easily rename the layout.

Copy Layout from another Drawing

Do you have a layout already set up in another drawing, and do you want to use it in this drawing? You can import a layout from any drawing file:

1. Continue using the `1-12_Using Layout Tabs.dwg` file.

2. *Right-click* on any Layout Tab in this drawing file and select the **From Template** command.

3. Navigate to the location of the file containing the completed layout and change the **Files of Type** to **.DWG** if needed.

4. Select the borders file, which contains all the standard layouts for this course, and click **Open**.

    `...\ACAD_TipsTechniques\Exercise_Files\Chapter 01\refs\Agency XYZ Borders.dwg`

5. Using the **Insert Layouts** dialog, select the **B-Border XYZ Sheet** layout and click **OK** to import the layout.

In the next section, we will learn to use some of the selection tools "hidden" features.

Bonus commands

In this section, we will examine how to control some of the interface's more obscure aspects, such as the Status Bar, how to use the QUICK PROPERTIES dialog to easily edit some of those hard-to-find dimension settings, and how to minimize the appearance of the AutoCAD tooltips.

First, let's look at how to control the "blinky" Status Bar and what causes this problem.

Blinky Status Bar?

Do you ever experience a "blinky" Status Bar when working in AutoCAD? Yes, it can be very distracting and needs to be addressed when it occurs. This problem is caused by the COORDINATES display in the Status Bar when it is too full of other icons and settings.

To prevent this problem, turn OFF the COORDINATES display in the Status Bar:

1. Open the 1-13_Quick Properties.dwg file.

2. Using the Status Bar, *left-click* on the "hamburger" icon ≡ and select the **Coordinates** setting to turn OFF this option in the Status Bar.

    By default, the COORDINATES are turned ON in the Status Bar. Depending on each individual user's Status Bar and screen width, having the COORDINATES turned ON may exceed the minimum width and cause this small display issue. If you need the COORDINATES to display, use the **Dock above Status Bar** layout setting to gain more display space for your Status Bar.

    27.3091, 17.9362, 0.0000

    Figure 1.24: Default coordinates display

3. *Right-click* on any Layout Tab and select **Dock above Status Bar** to separate the display of the Status Bar and Layout Tabs.

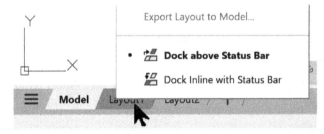

Figure 1.25: Status Bar and Layout Tabs control

In the next exercise, we will learn how to use the "hidden" QUICK PROPERTIES dialog.

Quick Properties

Users always ask me how to improve the availability of object properties for those objects that need to be edited on a regular basis. When I ask if they are using QUICK PROPERTIES, many users do not know what I am talking about.

Here are a couple of quick examples that demonstrate how you can use QUICK PROPERTIES to simplify your daily drafting workflow:

1. Continue using the 1-13_Quick Properties.dwg file.

2. Using the Status Bar, *left-click* on the "hamburger" icon ≡ and select **Quick Properties**.

3. This will turn ON the **Quick Properties** icon 🔳 in the Status Bar.

4. With Quick Properties enabled, *"hover"* on the new icon and *right-click* to access the **Quick Properties Settings**.

5. Using the **Drafting Settings** dialog, review the settings to control what properties are displayed and where in the **Quick Properties** dialog.

6. Click **CANCEL** to retain the default settings.

7. Select the DIMENSION object and notice the new **Properties** dialog that is displayed to the TOP RIGHT of the cursor.

8. This dialog may not display all the commonly used properties you need to review and modify, so let's change it to provide the properties we need for a TOLERANCE DIMENSION object.

Figure 1.26: Default Quick Properties dialog

Tolerance Dimensions

So, let's assume you need to edit the tolerance value in an existing dimension. When you *double left-click* on the dimension text, the entire text value is highlighted, which prevents you from just editing the tolerance values.

We need to add the TOLERANCE LIMITS and TOLERANCE PRECISION properties to the QUICK PROPERTIES dialog:

1. Continue using the 1-13_Quick Properties.dwg file.

2. Run the **CUI** command and use the **Customize User Interface** dialog to select the (**current**) workspace from the TOP LEFT **All Customization Files** panel.

3. Select the **Quick Properties** item in the list. This changes the display of the RIGHT panel in the dialog.

4. Using the RIGHT panel, select the **Rotated Dimension** object type and turn ON the **Tolerance limits** and **Tolerance precision** properties.

5. Click **Apply** and **OK** to save our changes and close the dialog.

6. Select the DIMENSION object and, using the **Quick Properties** dialog, modify the **Tolerance limit** values as needed.

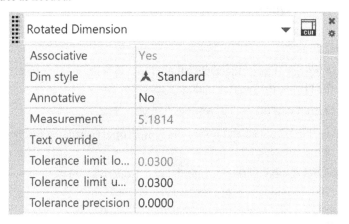

Figure 1.27: Modified Quick Properties dialog

In the next exercise, we will learn how to change the display of the default tooltips.

Dynamic tooltips

Use the TOOLTIPMERGE system variable to merge the "dynamic" tooltips together into a single tooltip. They are smaller and easier to read, and you won't see all those tooltips flying around your cursor!

TOOLTIPMERGE	
Combines drafting tooltips into a single tooltip.	
Type: Integer	
Saved In: Registry	
0 (default)	Drafting tooltips are separate
1	Drafting tooltips are merged

1. Open the 1-14_Tooltips.dwg.

2. Run the **Line** command and *left-click* in the view window to start the LINE object.

3. By default, the tooltips are all separated, as shown below.

4. Using the Command Line, key in **TOOLTIPMERGE** and press *Enter*. Key in the value of 1 to turn ON the merge effect.

5. Run the **Line** command again and *left-click* in the view window to start the LINE object.

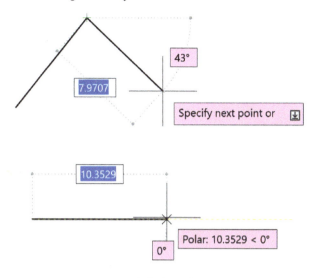

Figure 1.28: Default tooltip merge

Here are the merged dynamic tooltips:

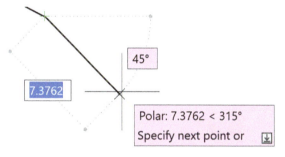

Figure 1.29: Merged dynamic tooltips

In these exercises, you learned some of the lesser-known interface options to customize your interface using the Status Bar, the Quick Properties dialog, and the Dynamic Input tooltips.

Summary

In this chapter, we examined how to use the newest features of the AutoCAD interface, including the new floating windows and customizing the Command Line's appearance and location. We learned how to use Dynamic Input more efficiently using key-ins that are not known to the majority of users. We reviewed how to use some useful keyboard shortcuts to simplify our daily workflow and how to assign custom keyboard shortcuts when needed. We learned how to make the most out of our selection commands and how to use more mouse controls, such as *drag and drop* for working with files.

Finally, we finished the chapter by reviewing some of the more obscure interface commands, such as Quick Properties and customized tooltips.

In the next chapter, we will look at how to take advantage of some of the advanced annotation features, including full control of text editing in the Mtext editor, controlling the appearance of our text objects, using fields for "smarter" text objects, and more controls for using annotation scale on objects.

2

Using More of the AutoCAD Interface

AutoCAD's interface is continuously improving, and with each new version, there are more options to consider. It's up to each user to assess what works best for their unique preferences.

In this chapter, you will learn about the additional tools for using the various *drag-and-drop* options in the new interface. We will further explore the different techniques used to control object selections and the associated display options for grips, selections, and groups of objects.

In this chapter, we will cover the following topics:

- *Drag-and-drop* everywhere
- Controlling the selection of everything
- Using grips more efficiently
- Using the bonus interface commands

By the end of this chapter, you will be able to customize the interface using more of the new tools and features.

Drag-and-drop

Now, let's look at how you can use the many *drag-and-drop* options to open files in AutoCAD. Make sure you are taking full advantage of all of these options to improve your daily workflow.

Open a file

First, let's look at how you can OPEN a drawing file using the following options.

Option 1 – using the AutoCAD title bar

1. Open the 2-0 Blank.dwg file.

2. Open a **Windows Explorer** dialog and navigate to the Exercise_Files folder:

    ..\ACAD_TipsTechniques\Exercise_Files\Chapter 02.

> **Note**
>
> Use the Windows shortcut key ⊞ + *E* to open the Windows Explorer dialog quickly.

3. Select the 2-1_Drag and Drop 1.dwg file and drag it from Windows Explorer onto the AutoCAD title bar to open a file.

4. Close this file using **X** on the **File** tab to prepare for the next option.

Option 2 – using the Command Line

1. Select the 2-1_Drag and Drop 1.dwg file and drag it from Windows Explorer onto the Command Line to open a file.

Figure 2.1: AutoCAD Command Line

2. Close this file using the **X** on the **File** tab to prepare for the next option.

Option 3 – using the AutoCAD "A" icon

1. Select the 2-1_Drag and Drop 1.dwg file and drag it from Windows Explorer onto the big red **"A"** at the LEFT side of the AutoCAD title bar to open a file.

Figure 2.2: Using the AutoCAD "A" icon in the title bar

2. Close this file using the **X** on the **File** tab to prepare for the next option.

Option 4 – using the AutoCAD Start tab

1. Select the 2-1_Drag and Drop 1.dwg file and drag it from Windows Explorer onto the AutoCAD **Start** tab to open a file.

Figure 2.3: The AutoCAD Start tab

2. Close this file using the **X** on the **File** tab to prepare for the next option.

Option 5 – using the desktop AutoCAD icon

If you use multiple versions of AutoCAD or multiple Autodesk applications, this tip will help you open files using the correct application.

1. Select the 2-1_Drag and Drop 1.dwg file and drag it from Windows Explorer onto the AutoCAD icon on your desktop or in Windows Explorer to open the file and run the application.

Figure 2.4: AutoCAD application icon

Once you finish this section, you can open a drawing file using different *drag-and-drop* options.

Insert a file with various results

In this section, we will learn how to insert a drawing file as a BLOCK using the *drag-and-drop* option.

As a block

1. Continue using the 2-1_Drag and Drop 1.dwg file.

2. Using the right mouse button, select the 2-1_Drag and Drop 2.dwg file and drag it from Windows Explorer into the drawing view window.

3. A menu appears, which gives you the following options:

   • **Insert Here**: This will insert the file as a BLOCK

   • **Open**: This will open the file

   • **Create Xref**: This will attach the file as an XREF

   • **Create Hyperlink Here**: This will insert the file as a hyperlink to the file

   • **Cancel**: This will cancel the *drag-and-drop* action

Insert Here
Open
Create Xref
Create Hyperlink Here
Cancel

Figure 2.5: Drag-and-drop right-click options

4.   Select the **Insert Here** option and *left-click* anywhere in the view window to place the file as a block. Press the *Enter* key three times to accept the default *X* scale factor, *Y* scale factor, and *rotation* angle.

Insert a text file

In this exercise, let's look at inserting a text file as a BLOCK using the *drag-and-drop* option.

As a block

1.   Continue using the 2-1_Drag and Drop 1.dwg file.
2.   Select the 2-1_General Notes.txt file and drag it from Windows Explorer into the drawing view window. All formatting from the original text file is transferred and placed as MTEXT using the current text style.

Insert a PDF file

In this exercise, let's look at inserting a PDF file as an image using the *drag-and-drop* option:

1.   Continue using the 2-1_Drag and Drop 1.dwg file.
2.   Select the 2-1_AutoCAD Shortcuts.pdf file and drag it from Windows Explorer into the drawing window.
3.   If the PDF has multiple pages, as this one does, you are prompted for what page to insert. Enter page number 5.
4.   *Left-click* in the drawing window to identify the insertion point for the PDF image, press the *Enter* key to accept the scale of 1, and press *Enter* again to accept the rotation of 0.
5.   The resulting PDF image has been attached as a **PDF Underlay** and will appear in the **References** dialog and any other reference files.

After finishing these exercises, you can utilize several *drag-and-drop* functions to manage different file formats, including inserting files, texts, and PDFs.

Working with selections

This section will teach us to use some of the most important features of creating selection sets. Becoming efficient with creating and using selection sets is crucial to improving your overall efficiency with AutoCAD.

Control the Lasso

First, let's learn how to use the new Lasso selection method to gather up those objects and corral them into place quickly! Lasso is available, by default, when you *left-click* and *drag* the cursor around existing objects:

1. Open the `2-2_Selecting Objects.dwg` file.

2. Using the In-Canvas View Controls, restore the **Custom Model Views | 1-Using LASSO** to apply the named view to the current viewport.

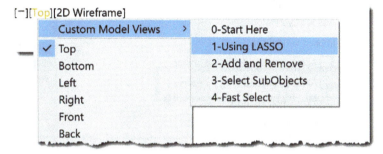

Figure 2.6: Custom Model Views | 1-Using Lasso

3. *Left-click and drag* the cursor around the objects to select.

4. Release the left mouse button to complete the selection. Use any command to modify the selection.

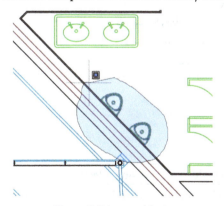

Figure 2.7: Lasso objects

> **Note**
>
> Remember, when selecting objects, *drag right-to-left* (counter-clockwise) to use an implied "green" crossing (overlap) selection, and *drag left-to-right* (clockwise) to use an implied "blue" window (inside) selection.

Many users find the Lasso selection method too intrusive to use regularly. You can disable it using **Options | Selection | Selection Modes**.

I prefer to disable it by default and assign a shortcut key to enable it when needed, as demonstrated in the following steps:

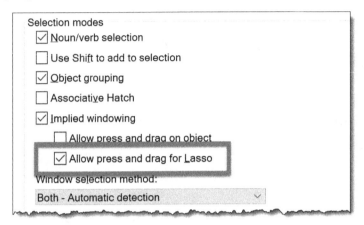

Figure 2.8: Lasso options

Assigning Lasso to a shortcut key

Before assigning a shortcut key, you must learn how to enable and disable the Lasso feature using a system variable. Using the following system variable, we can change PICKAUTO to 1 to disable the Lasso feature:

PICKAUTO	
Controls automatic windowing selection options for selecting objects.	
Type: Bitcode	
Saved In: Registry	
Initial Value: 5	
0	Turns off the automatic window and crossing selection.
1	A Window or Crossing selection begins even if the cursor is not directly over an object when you *Left-click and drag*.
2	A window or crossing selection begins even if the cursor is directly over an object. This option applies only when you select objects before you start a command using the *Left-Click and drag* method.
4	Begins a window or crossing lasso selection if the cursor is not directly over an object when you click and drag. *Using Left-Click and drag* will begin a free-form fence selection if the Fence option is active.

In the next exercise, you will learn how to make a shortcut key to toggle the Lasso selection setting ON and OFF.

Creating a Lasso shortcut key

There are several methods for creating a custom command in AutoCAD. For this example, we will record an ACTION MACRO and assign it to a shortcut key.

First, we must record the macros to turn the Lasso setting ON and OFF using the **Options** dialog.

1. Continue using the 2-2_Selecting Objects.dwg file.

Macro Lasso-OFF

1. Using the **Manage** ribbon and the **Action Recorder** panel, select the **Record** command.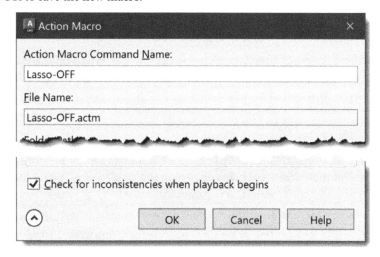
2. Using the Command Line, key in the PICKAUTO system variable.
3. Key in the new 0 value to turn OFF the Lasso selection method.
4. Using the ribbon, select the **Stop** command.
5. Using the **Action Macro** dialog, key in the name Lasso-OFF.
6. Click **OK** to save the new macro.

Figure 2.9: Lasso-OFF Macro

Macro Lasso-ON

1. Using the ribbon, select the **Record** command.

2. Using the Command Line, key in the PICKAUTO system variable.

3. Key in the new 5 value to turn ON the Lasso selection method ON.

4. Using the ribbon, select the **Stop** command.

5. Using the **Action Macro** dialog, key in the name Lasso-ON.

6. Click **OK** to save the new macro.

Next, we will learn to assign the two new macros to custom shortcut keys.

Creating shortcut keys

CUI	Command Locations
Ribbon	Manage | Customization | User Interface
Command Line	CUI (CUI)

Let's create a new shortcut keys for our custom commands for Lasso-ON and Lasso-OFF.

1. Continue using the 2-2_Selecting Objects.dwg file.

2. Using the Command Line, key in the CUI command.

3. Using the **Customize User Interface** dialog, select the (**current**) workspace from the TOP LEFT panel.

4. Expand the **Keyboard Shortcuts** item in the list and expand **Shortcut Keys**.

> **Note**
> At the time of this printing, ACTION MACROS do not display in the **CUI | Command List**. To overcome this, we must create our own commands and assign the ACTION MACROS to the new commands using the **Create a new command** button.

5. Using the LOWER LEFT portion of the dialog, locate the **Command List** section, and click the **Create a new command** button. 🌟

6. Using the **Properties** section, located in the LOWER LEFT portion of the dialog, change the **Name** field from Command1 to Lasso-ON.

7. Change the **Macro** field from ^C^C to ^C^C_Lasso-ON for Lasso-ON and from ^C^C to ^C^C_Lasso-OFF for Lasso-OFF.

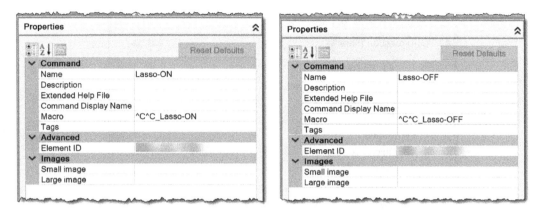

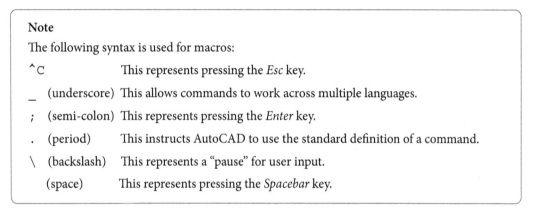

Figure 2.10: Custom commands-CUI

> **Note**
>
> The following syntax is used for macros:
>
> ^C   This represents pressing the *Esc* key.
>
> _   (underscore) This allows commands to work across multiple languages.
>
> ;   (semi-colon) This represents pressing the *Enter* key.
>
> .   (period)   This instructs AutoCAD to use the standard definition of a command.
>
> \   (backslash)   This represents a "pause" for user input.
>
> (space)   This represents pressing the *Spacebar* key.

Now that the new Lasso commands are available in the CUI dialog, we can assign them to keyboard shortcuts.

1. Using the **Command List** search field, located in the LOWER LEFT section of the dialog, *drag and drop* the **Lasso-ON** command into the **Keyboard Shortcuts | Shortcut Keys** list.

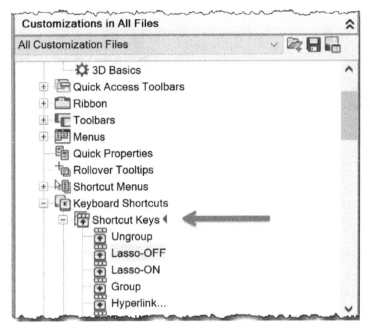

Figure 2.11: Drag-and-drop to Shortcut Key

2.   Repeat *Step 5-8* from the previous exercise to create the **Lasso-OFF** command.

3.   Click **Apply** to save these changes.

Now, we will assign specific shortcut keys to these commands.

4.   Using the **Keyboard Shortcuts | Shortcut Keys** list, select **Lasso-ON**.

5.   Using the **Properties** section, located in the LOWER RIGHT portion of the dialog, select the **Key(s)** value, and click the **Browse** button to open the **Shortcut Keys** dialog.

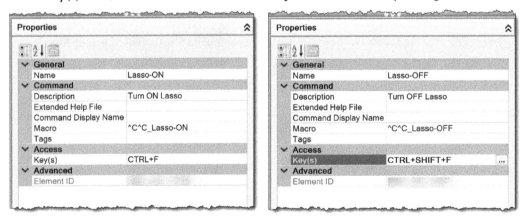

Figure 2.12: The Lasso ON/OFF shortcut key properties

> **Note**
>
> When defining the actual shortcut keys, select the **Browse ...** button and key in the desired shortcut keys. If you do not get the correct keys when using the keyboard to set the shortcut keys, verify that the *Caps Lock* key is not activated on the keyboard.

6. Click **Apply** to save these changes.

7. Click **OK** to close the dialog.

8. Test both keyboard shortcuts for **Lasso ON** (*Ctrl + F*) and **Lasso OFF** (*Ctrl + Shift + F*).

> **Note**
>
> There is also a FENCE option available during any selection operation. Key in F in the Command Line at the **Select Objects** prompt and *left-click* and *drag* a line through the objects you want to select. Repeat using the left mouse button to add additional fence selections.

In the next section, we will learn how to add and remove objects from an existing selection set.

Adding and removing objects

When working with an existing selection set, do you always get the perfect selection? Of course not! Do you use all the shortcut methods available for removing and adding objects to a selection set? Let's find out:

1. Continue using the `2-2_Selecting Objects.dwg` file.

2. Using the In-Canvas View Controls, restore the **Custom Model Views | 2-Add and Remove** to apply the named view to the current viewport.

3. *Drag* a **Window** or **Crossing** selection around multiple objects in the drawing to select one of the elevator bays.

4. Use the *Shift* key to add and remove objects from the selection set if you accidentally miss or select too many objects.

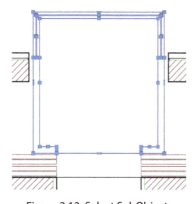

Figure 2.13: Select SubObjects

Select SubObjects

In this exercise, we want to remove the BOTTOM line that is displayed at the elevator door opening:

1. Continue using the 2-2_Selecting Objects.dwg file.

2. Using the In-Canvas View Controls, restore the **Custom Model Views | 3-Select SubObjects** to apply the named view to the current viewport.

3. Select the exterior RECTANGLE object around the exterior of the elevators.

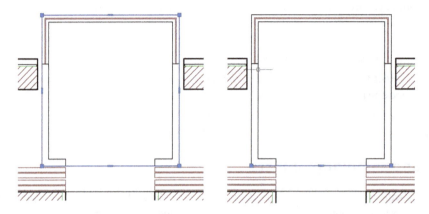

Figure 2.14: Selecting SubObjects

As you can see, the polyline is closed and cannot be deleted without deleting the entire polyline. I want to delete just the BOTTOM edge of the rectangle shape to allow for the door opening. Most users would use the EXPLODE command to break up the closed polyline and then use the DELETE command to remove the BOTTOM line. However, there is a quicker way! You can use the SubObject selection feature to select one object in the polyline to remove it.

4. Use the *Ctrl* key to select the **BOTTOM edge** as a SubObject of the closed polyline. Then, with just the bottom edge selected, use the *Delete* key to delete the selected line. Using this shortcut, you can avoid using additional commands such as EXPLODE, ERASE, and JOIN.

In the next exercise, we will learn how to control the display of selected objects in AutoCAD.

Glowing selection effect

When selecting objects in AutoCAD, the objects are displayed using the "glowing" effect by default. If you do not like the "glowing" effect of the selected objects and you prefer the old "dashed" effect of the selected objects, you can modify the selection dynamic effects, and disable the "glowing" effect using the SELECTIONEFFECT system variable.

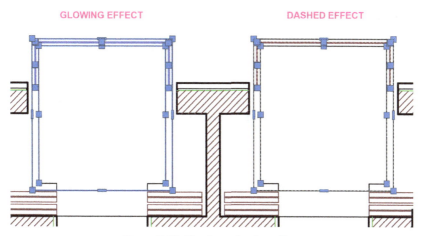

GLOWING EFFECT   DASHED EFFECT

Figure 2.15: Glowing selection effect

SELECTIONEFFECT	
Specifies the visual effect used when objects are selected.	
Saved In: Registry	
Type: Integer	
0	Dashed lines
1 (default)	When hardware acceleration is on, displays a glowing line highlighting effect

In the next exercise, we will learn how to use the Express Tool, **Fast Select**, to easily select all "touching" objects in a drawing.

Fast Select (Express Tool)

Selecting objects can be tricky sometimes when trying to avoid or select multiple objects. Use the Express Tool, **Fast Select**, to select all objects that "touch" a selected object. This command will select lines, polylines, circles, arcs, attribute definitions, text, mtext, ellipses, and images. To use this command, follow these steps:

FS	Command Locations
Command Line	FS

1. Continue using the 2-2_Selecting Objects.dwg file.

2. Using the In-Canvas View Controls, restore the **Custom Model Views | 4-Fast Select** to apply the named view to the current viewport.

3. Using the Command Line, key in FS to run the **Fast Select** command.

4.   Select the perimeter lines of lots 9-15 at **P1** and all the connected lines are selected automatically regardless of their properties.

5.   Use the *Esc* key to clear the current selection set.

Note

Use the key in ' FS to use this command "transparently" inside of another command at the SELECT OBJECTS prompt.

The behavior of the **FS** command is controlled using the **FSMODE** command:

FSMODE	Command Locations
Command Line	FSMODE
ON	Selects all the objects that touch the selected object and any objects that are touching those objects.
OFF	Selects only the objects that are touching the selected object.

1.   Using the Command Line, key in FSMODE and key in ON to change the functionality of the **Fast Select** command. With FSMODE set to ON, all the objects that are touching the selected object, and all objects that are touching those objects, are selected.

2.   Using the Command Line, key in FS to run the **Fast Select** command again.

3.   Select the previously selected line at **P1**, and all the connected lines, and their touching objects, are selected automatically regardless of their properties.

Figure 2.16: The Fast Select results

4.   Use the *Esc* key to clear the current selection set.

After completing this exercise, you can fully take advantage of the selection tools provided in AutoCAD. With these additional commands, you can now better control your selection sets and common manipulations.

In the next section, we will take a look at how to use the OBJECT ISOLATION to control the view of objects in a drawing.

Using object isolation

When working in drawings with thousands of objects, using layers to "filter" your view display may not be enough. You can use the OBJECT ISOLATE tools to filter just what you need and minimize your view extent and your frustration. Working with a minimum number of objects displayed not only speeds up AutoCAD but also improves the tasks you need to perform.

Isolate objects

Use the ISOLATE OBJECTS command to filter the display of selected objects and turn off all objects not in the current selection set.

1. Continue using the 2-2_Selecting Objects.dwg file.
2. Select several objects in the view window and, using the Status Bar, *left-click* on the **Isolate Objects** icon ⌐₀ to access the **Isolate Objects** command.

Only the selected objects are displayed in the drawing. All other objects are temporarily "hidden" from the view. The ISOLATE OBJECTS icon also changes to a blueish appearance, indicating that some objects in the file are hidden. 🔲

With objects isolated in the drawing, you have the following command options available when you *left-click* on the ISOLATE OBJECTS command in the Status Bar.

- **Add Additional Objects**: Create a new object selection set to isolate
- **Hide Objects**: This will hide additional objects in the view
- **End Object Isolation**: Use this to restore all objects in the view

With a selection set active, you can also access the ISOLATE OBJECTS command using the *right-click* menu:

1. Continue using the 2-2_Selecting Objects.dwg file.
2. *Left-click* on the **Isolate Objects** command ⌐₀ in the Status Bar to access the **End Object Isolation** command and restore all objects in the view.
3. Again, select several objects in the view window and *right-click* to access the **Isolate | Isolate Objects** command.
4. *Right-click* again to access the **Isolate | Isolate Objects, Isolate | Hide Objects**, or **Isolate | End Object Isolation** command.

In the next section, we will look at how to use Grips in AutoCAD more efficiently.

Using Grips efficiently

Using Grips in AutoCAD can speed up simple object editing without any command. First, let's review the basic GRIP options for the common object types for those who don't use Grips now.

Using Common Grip commands

Grips in AutoCAD provides access to common editing commands such as STRETCH, MOVE, ROTATE, MIRROR, and SCALE. Most users should know these simple GRIP functions, but just in case you don't, let me demonstrate them quickly in the next exercise.

Using Common Grips

In this exercise, we will use Grips to do a simple COPY and ROTATE as an example:

1. Open the 2-3_Using GRIPS.dwg file.
2. Using the In-Canvas View Controls, restore the **Custom Model Views | 1-Basic Grips** named view.
3. Select the RECTANGLE object and notice the "blue" Grips that display at the ENDPOINT and MIDPOINT locations on the object.
4. Select one of the corner Grips (blue square grip) and look at the Command Line.
5. When selecting a grip, you are automatically placed in the **Stretch** command. Use the *Spacebar* to cycle to the next command, which is **Move**. Use the **Cursor Badges** to keep track of which command mode you are using. ⊕

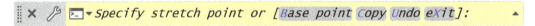

Figure 2.17: Grips common commands

6. Press the *Spacebar* again to cycle to the next command, which is **Rotate**. ⊙
7. Press the *Spacebar* again to cycle to the next command, the **Scale** command. ⬛
8. Press the *Spacebar* again to cycle to the next command, the **Mirror** command.
9. Last, press the *Spacebar* again to cycle back to the first command, the **Stretch** command.
10. Use the *Esc* key to cancel the command and to clear the selection set.

    Standard practice is to cycle through the **Stretch**, **Move**, **Rotate**, **Scale**, and **Mirror** commands and back to Stretch again using the *Spacebar*. However, you can access them directly by entering their abbreviations, as shown here:

    -   ST – Stretch ⬀
    -   MO – Move ⊕

- RO – Rotate ↻
- SC – Scale ⬜
- MI – Mirror ⚠

> **Note**
>
> You can also access the GRIP commands using a *right-click* menu.

11. Select the RECTANGLE object and select one of the corner Grips (blue square grip).

12. By default, you are in the **Stretch** command. Key in RO in the Command Line to skip immediately to the **Rotate** command.

13. Use the *Esc* key to cancel the command and to clear the selection set.

This is a quick shortcut that is handy to know.

> **Note**
>
> The COPY command is not a grip mode but can be selected as an option within any of the other GRIP commands.

> **Note**
>
> Try using the *Ctrl* key when using the GRIP command. For example, if you press *Ctrl* during the ROTATE COPY command, the angle is repeated from the previous ROTATE COPY. If you release the *Ctrl* key, the angle is free to copy at any angle.

Let's give that a try in the next exercise:

1. Continue using the 2-3_Using GRIPS.dwg file.

2. Select the SQUARE object and select any GRIP on the object.

3. Use the *Spacebar* to toggle to the **Move** command and key in C to make the MOVE action a **Move-Copy** of the SQUARE object.

4. Drag the SQUARE object to the RIGHT, and key in the value of 1 unit, and *left-click* to complete the **Move-Copy**.

5. While still in this command, hold down the *Ctrl* key and drag the SQUARE object to the RIGHT until it locks into the 1-unit distance.

6. *Left-click* to accept the new **Move-Copy**. Release *Ctrl* when you no longer want to use the snap distance for additional copies.

> **Note**
>
> Try using this *Ctrl* key option for any of the common GRIP commands.

Performing simple modifications on AutoCAD objects using these common GRIP commands can save time and improve your overall editing process.

Now, let's look at what else these Grips provide besides the common commands we just reviewed.

Using Multi-Functional OBJECT Grips

In the following exercises, we will learn to use additional GRIP commands often overlooked by users. To access these multi-functional grips, you must *hover* over a grip, not select or *left-click* on a grip.

Lines

First, let's look at an example of using these multi-functional grips to STRETCH a line and maintain the existing angle of the line:

1. Continue using the 2-3_Using GRIPS.DWG file.

2. Using the Status Bar, turn OFF the **Dynamic Input** setting ⁺⁻ for this first example.

3. Select the LINE object and review the grips (blue squares) at the ENDPOINT and MIDPOINT locations.

4. *Hover* over either ENDPOINT grip (blue square) to display the multi-functional pop-up menu, which gives you access to the **Stretch** and **Lengthen** commands.

5. Select the **Lengthen** command from the pop-up menu to extend the existing LINE without modifying the angle of the line.

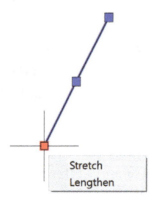

Figure 2.18: Lengthen Grip

6. *Left-click* anywhere along the LINE object to define the length change and to complete the command.

7. Using the Status Bar, turn ON the **Dynamic Input** setting ⊞ for the next example.

8. Again, select the LINE object and *hover* over either ENDPOINT grip to access the **Stretch** and **Lengthen** commands. Note the additional dynamic dimensions available to visually control the length modification.

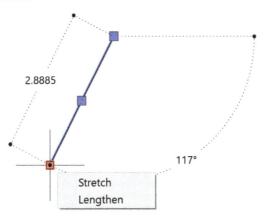

Figure 2.19: Lengthen Grip with Dynamic Input

Polylines

Next, let's look at how to modify a POLYLINE object and quickly insert a new vertex using a multi-functional grip rather than an individual AutoCAD command:

1. Continue using the 2-3_Using GRIPS.DWG file.

2. Select the RECTANGLE polyline object and *hover* over any corner grip (blue square) to access the multi-functional pop-up menu for the **Stretch**, **Add Vertex**, and **Remove Vertex** commands.

3. Select the **Add Vertex** command from the pop-up menu to add a new vertex to the "next" edge of the rectangle. The new vertex is added between the "hover" grip and the next grip in the existing object's vertices.

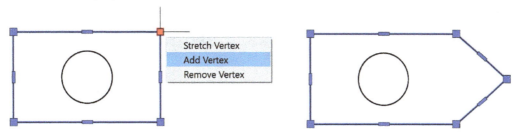

Figure 2.20: Add Vertex Grip

4. *Left-click* in the view window to add another vertex to this POLYLINE shape. This is much easier than using the **Pedit** command to add a vertex, isn't it?

5. Use the *Esc* key on the keyboard to cancel all commands and clear all grips.

Utilizing these "hover" grip commands can provide greater flexibility and efficiency compared to using the identical command options available in each individual AutoCAD command.

Try out the following object GRIP options for these object types.

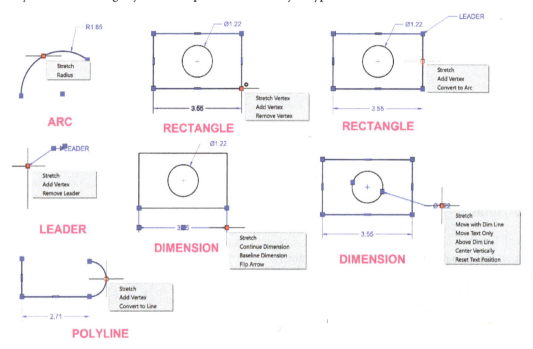

Figure 2.21: Additional GRIP commands

Selecting multiple GRIPS

When using grips, you can also select multiple grips on an object to better control the grip editing:

1. Continue using the 2-3_Using GRIPS.DWG file.

2. Select the RECTANGLE polyline object and use the *Shift* key to select two corner grips.

3. When both grips are selected (red squares), select either grip to edit the rectangle. With multiple grips selected, you lock their relationship to each other, allowing you to edit the RECTANGLE object differently.

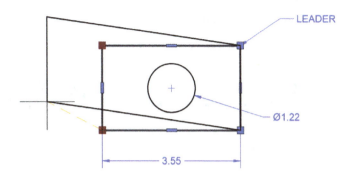

Figure 2.22: Select Multiple Grips

Customize the GRIP appearance

Last, let's look at how you can customize the appearance of GRIPS in your AutoCAD.

Grip size

1. Continue using the 2-3_Using GRIPS.DWG file.

2. *Right-click* anywhere in the view window and select the **Options** command.

3. Using the **Options** dialog, select the **Selection** tab.

4. Using the **Grip Size** slider bar, adjust the size of your grips as needed. The default size is **5**. You can also use the following system variable to control the grip size.

GRIPSIZE	
Sets the size of the grip box, in device independent pixels.	
Type: Integer	
Saved in: Registry	
1-255	Defines the size of the GRIP icon.
5 (default)	

I would like to modify the color of my grips to make them stand out better against the default dark color scheme and background.

Grip colors

Modify the colors of your grips to suit your view background color and the commonly used colors in your CAD standards:

1. Continue using the file 2-3_Using GRIPS.DWG.

2. *Right-click* anywhere in the view window and select the **Options** command.

3.   Using the **Options** dialog, select the **Selection** tab.

4.   Select the **Grip Colors** button and using the **Grip Colors** dialog, modify the colors to your personal preference.

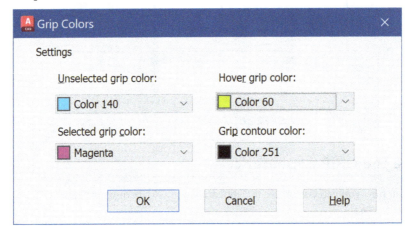

Figure 2.23: Grip colors

5.   Click **OK** to save the changes and click **OK** again to close the **Options** dialog.

As with other GRIP settings, the GRIP colors can be defined using the following system variables:

GRIPCOLOR	
Controls the color of unselected grips. The valid range is 1 to 255.	
Type: Integer	
Saved in: Registry	
1-255	Defines the unselected GRIP color.
150 (default)	

GRIPHOVER	
Controls the fill color of an unselected grip when the cursor pauses over it. The valid range is 1 to 255.	
Type: Integer	
Saved in: Registry	
1-255	Defines the hover GRIP color.
150 (default)	

GRIPHOT	
Controls the color of selected grips. The valid range is 1 to 255. Type: Integer Saved in: Registry	
1-255	Defines the selected GRIP color.
12 (default)	

GRIPCONTOUR	
Controls the color of the grip outline. The valid range is 1 to 255. Type: Integer Saved in: Registry	
1-255	Defines the grip outline color.
251 (default)	

Grip display settings

The remaining GRIP options include the ability to disable the use of Grips altogether:

GRIPS	
Controls the display of grips on selected objects. Type: Integer Saved in: Registry	
0	Hides grip display.
1	Displays grips
2 (default)	Displays additional midpoint grips on polyline segments.

Show Grips in blocks

Many times, displaying GRIPS for all objects in a block can be overwhelming. Using this setting, you can control whether the GRIPS displayed for block objects display only the block INSERTION location or all grips for the objects in the block.

Grip display limit

You can control how many grips are displayed in a multiple object selection set by defining the **Object Selection Limit for display of grips** setting, as shown in the following image. By default, AutoCAD will not display more than 100 grips in a selection set:

1. Continue using the 2-3_Using GRIPS.dwg file.

2. Using the In-Canvas View Controls, restore the **Custom Model Views | 1-Grip Limits** named view.

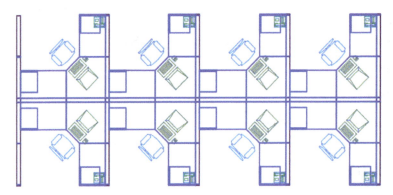

Figure 2.24: Grips limit 100

3. *Right-click* anywhere in the view window and select the **Options** command.

4. Using the **Options** dialog, select the **Selection** tab.

5. Using the **Grips** section of the tab, locate the **Object selection limit for display of grips** setting and modify the value to 200.

6. Click the **OK** button to close the dialog.

7. Select several objects again and review the changes to this setting.

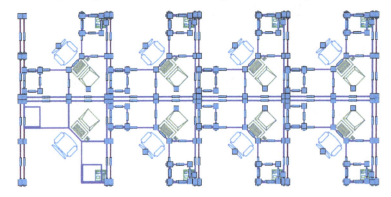

Figure 2.25: Grips limit 200

After completing this section, you can more fully utilize and customize your personal use of GRIPS in AutoCAD. Continue to explore what you can do with GRIPS to improve your editing efficiency.

Next, let's learn how to use the GROUP commands for more than you have imagined.

Bonus commands

Let's look at the GROUP command options often overlooked in AutoCAD. The first is the GROUP command, which allows you to create on-the-fly block-like objects. So, when would this command be most useful? When editing multiple objects in a drawing, you can create a selection set of objects, and instead of creating a BLOCK of those objects, use the GROUP command to group the objects together temporarily or permanently with or without a name.

So, where do you find the GROUP command?

GROUP	Command Locations
Ribbon	Home | Groups | Group
Command Line	GROUP (GRO)
	-GROUP to display Command Line options

The GROUP command

In this exercise, we will learn how to use group objects to simplify multiple object manipulation.

1. Open the 2-4_Bonus INTERFACE Commands.dwg file.
2. Select the three SQUARE objects in the TOP row.
3. Using the **Home** ribbon and the **Groups** panel, select the **Group** command 🖼 and press *Enter* to complete the command.
4. *Hover* over the TOP row objects and note how the three SQUARE objects now highlight as a single object, not three individual objects. They are now considered a single GROUP object.

> **Note**
> If the three SQUARE objects do not highlight as a GROUP, your GROUP toggle may be turned OFF. Use *Ctrl + Shift + A* to toggle the GROUP setting ON and OFF.

5. Select the new GROUP object, *right-click* to access the *right-click* shortcut menu, and select the **Properties** command.
6. Using the **Properties** dialog, notice that the three polylines are now considered a single GROUP object.

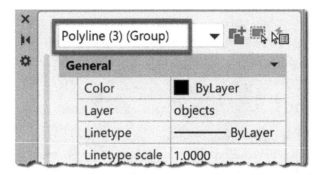

Figure 2.26: GROUP Properties

7. Repeat *Steps 3-4* and create new GROUPS for the MIDDLE row, BOTTOM row, and ALL nine square objects.

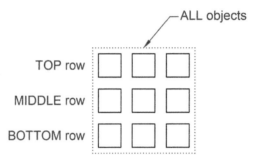

Figure 2.27: Create groups

8. Using the newly created groups, select one of the three squares in the BOTTOM row. All nine objects are selected using the ALL Objects group, not just the objects in the BOTTOM row.

    So, how do we work with the embedded BOTTOM row group?

9. Select the GRIP (blue square) for the BOTTOM row group (three squares) and use the **Move** command to move the BOTTOM row DOWN 2-units.

10. The BOTTOM row group is moved as a single object but is still a part of the ALL Objects group.

So, even with embedded GROUPS, you can easily move multiple objects as needed.

Depending on which object you select in the GROUP, the related GRIPS will display to enable you to use these GRIPS to perform the common Stretch, Move, Rotate, and Scale grip commands, as discussed earlier in this chapter.

Group Bounding Box

When you select a GROUP of objects, the boundary will display by default. You can control the boundary display by selecting the GROUPS ribbon drop-down or expanded panel and toggling the **Group Bounding Box** setting ON or OFF, or using the following system variable.

GROUPDISPLAYMODE	
Controls how grips display when GROUPS are enabled and an object in a group is selected.	
Type: Integer	
Saved in: Registry	
0	Displays grips on all objects in the selected group.
1	Displays a single grip at the center of the grouped objects.
2 (default)	Displays a bounding box around the grouped objects, and displays a single grip at the center.

The UNGROUP command

Next, we want to learn how to ungroup these objects.

1.   Continue using the 2-4_Bonus INTERFACE Commands.dwg file.

2.   Select one of the SQUARE objects in the TOP row.

3.   Using the **Home** ribbon and the **Groups** panel, select the **Ungroup** command.

4.   *Hover* over one of the squares in the TOP row and notice that the group definition has been removed.

Removing the GROUP definition is not always the best solution when you want to manipulate the GROUP objects individually. I recommend using the GROUP SELECTION mode, ON or OFF, to toggle the ability to manipulate the objects as a GROUP or as individual objects.

GROUP selection mode

In this exercise, we will control the recognition of GROUP objects using the **Group Selection** ON and OFF toggle.

1.   Continue using the 2-4_Bonus INTERFACE Commands.dwg file.

2.   Using the **Home** ribbon and the **Groups** panel, select the **Group Selection On/Off** command to turn the **Group Selection** OFF. Notice in the Command Line there is no indication that this command has been selected. The only way to tell whether Group Selection is ON or OFF is to check the ribbon button or *hover* over an object.

3.   Using the ribbon, if the **Group Selection ON/OFF** button is "blue" in color, the selection mode is **ON**. If the button is "gray" in color, the selection mode is **OFF**.

4. You can also *hover* over an object and review what objects highlight to determine whether **Group Selection** is **ON** or **OFF**.

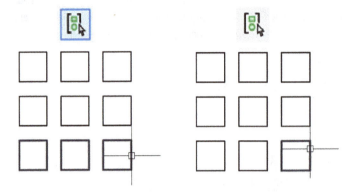

Figure 2.28: GROUP Selection ON/OFF

5. *Hover* over a SQUARE object in the BOTTOM row and notice that only an individual object is highlighted.

6. Using the **Home** ribbon and the **Groups** panel, select the **Group Selection ON/OFF** command again to turn **Group Selection** ON.

7. *Hover* over the SQUARE object in the BOTTOM row again, and notice the group of three squares is highlighted.

The **Group Selection ON/OFF** command toggles the group recognition ON and OFF.

> **Note**
>
> You can use the *Ctrl + Shift + A* keyboard shortcut to also toggle the **Group Selection ON/OFF** command.

I also prefer to assign a keyboard shortcut key to the GROUP and UNGROUP commands. These shortcut keys make it much easier to work with these commands on the fly.

Group edit

Unless your selection operations are perfect and you never miss-select an object, you must learn how to edit an existing group to add and remove objects. This can be accomplished by using the GROUP EDIT command:

1. Continue using the 2-4_Bonus INTERFACE Commands.dwg file.

2. Using the **Home** ribbon and the **Groups** panel, select the **Group Edit** command and select the TOP row group.

3. Using the Command Line, key in R to remove objects from the group.

4. Select the TOP LEFT SQUARE object to remove it from the group and press the *Enter* key to complete the command.

   We can also use a *right-click* shortcut menu to access the GROUP EDIT command.

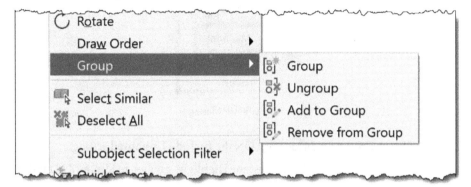

Figure 2.29: Using the Group menu

5. Use the *Esc* key on the keyboard to cancel any running commands.

6. Select the TOP RIGHT SQUARE object and *right-click* to access the **GROUP | Add to Group** command.

7. Select the TOP RIGHT SQUARE object to add it back into the group.

8. Use the *Enter* key to complete the command.

Using either of these two command methods will give you access to the GROUP commands. Choose the interface method that is most comfortable for you.

If you find you make numerous groups and maintain them in your drawing files, you will want to assign logical names to them so they are easier to identify using the Group Manager.

Using the Group Manager

The Group Manager allows you to assign logical names to your saved groups:

1. Continue using the 2-4_Bonus INTERFACE Commands.dwg file.

2. Using the **Home** ribbon and the **Groups** panel, select the expanded **Group** panel, and select the **Group Manager** command.

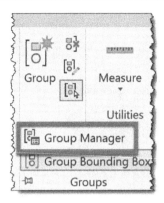

Figure 2.30: GROUP Manager

3. Using the **Object Grouping** dialog, turn ON the **Include Unnamed** setting.

4. Select the first group in the list (***A1**) and use the **Highlight <** button to identify it in the drawing.

5. Use the **Continue** button to close the **Highlight** dialog.

6. With this group (***A1**) still selected in the list, use the **Group Name** field and rename this group to **TOP**.

Note

The RENAME command in the Object Grouping dialog does not allow spaces, including variations or combinations of spaces.

---

7. Click the **Rename** button to apply the new name to the selected group.

8. Click **OK** to close the dialog.

To ensure your GROUPS are well organized, utilize the Group Manager and assign them logical names. Additionally, consider creating custom shortcut keys to automate the use of GROUP commands for better efficiency.

Using shortcut keys to Group and Ungroup

As you may have noticed, using groups is a convenient method to manage several objects at once. However, I personally utilize this function frequently enough that I use custom shortcut keys as a faster approach to access the GROUP and UNGROUP commands.

In this exercise, we will assign keyboard shortcuts to speed up the use of these commands.

First, let's assign a shortcut key for the GROUP command:

1. Continue using the 2-4_Bonus INTERFACE Commands.dwg file.

2. Using the Command Line dialog, key in the CUI command, and using the **Customize User Interface** dialog, select the (**current**) workspace.

3. Expand the **Keyboard Shortcuts** item and expand the **Shortcut Keys** item.

4. Using the **Command List** panel, located in the LOWER LEFT, select the **Search** field and key in GROUP to filter the list of commands.

5. Select the GROUP command and *left-click and drag* it into the **Keyboard Shortcuts** list.

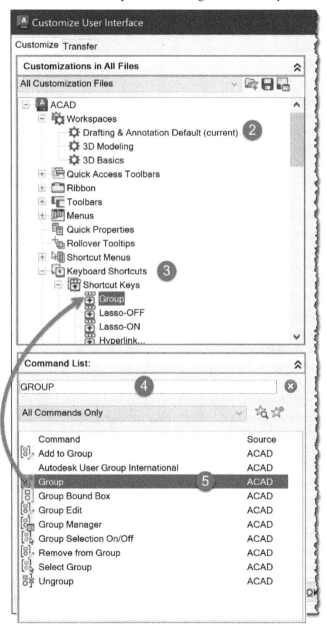

Figure 2.31: Create a GROUP shortcut key

6. Select the **Group** command in the **Shortcut Keys** list.

7. Using the **Properties** panel in this dialog, located in the LOWER RIGHT, select the **Access |
    Keys | ...** to open the **Shortcut Keys** dialog.

8. Key in CTRL + G for the new shortcut key and click **OK** to close the dialog.

9. Click **Apply** to save this new shortcut key.

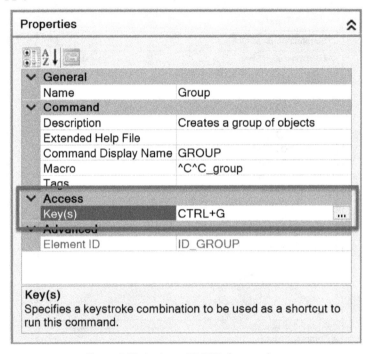

Figure 2.32: Assign a GROUP shortcut key

Now, let's assign a shortcut key for the UNGROUP command:

1. Using the **Command List** panel, located in the LOWER LEFT, select the SEARCH field and
    key in UNGROUP to filter the list of commands.

2. Select the UNGROUP command and *left-click and drag* it into the **Keyboard Shortcuts** list.

3. Select the UNGROUP command in the **Shortcut Keys** list.

4. Using the **Properties** panel in this dialog, located in the LOWER RIGHT, select **Access | Keys
    | ...** to open the **Shortcut Keys** dialog.

5. Key in CTRL + U for the new shortcut key and click **OK** to close the dialog.

6. Click **Apply** to SAVE this new shortcut key, and click **OK** to close the dialog.

Now, let's try the new shortcut keys for the GROUP and UNGROUP commands:

1. Continue using the 2-4_Bonus INTERFACE Commands.dwg file.

2. Select all nine SQUARE objects, and use the *Ctrl + G* shortcut key to make them into a group.

3. Using the **Home** ribbon and the **Groups** panel, select the expanded **Group** panel, and verify the **Group Selection** command is **ON** (blue).

4. *Hover* over one of the SQUARE objects to verify that all nine objects are highlighted as a group.

5. Select the new GROUP of objects and use the shortcut key *Ctrl + U* to ungroup these objects.

6. Press *Enter* key to accept this change.

7. *Hover* over one of the SQUARE objects to verify that only one object highlights and the GROUP definition has been removed.

Using these shortcut keys will give you much quicker access to these extremely useful tools and encourage you to use them more often.

Summary

In this chapter, we examined how to use the various *drag-and-drop* opportunities in AutoCAD and how to be more efficient using the selection methods provided. We also learned how to use grips on various object types to perform common manipulations such as STRETCH, MOVE, ROTATION, SCALE, and MIRROR.

We finished this chapter by learning how to use the GROUP and UNGROUP commands with custom keyboard shortcuts to make these commands more accessible and user-friendly while promoting their increased usage.

In the next chapter, we will examine how to take advantage of some of the advanced annotation features, including full control of text editing in the MTEXT Editor and the appearance of our text objects, using fields for "smarter" text objects, and more controls for using annotation scale on objects.

3
Taking Advantage of Annotation

In this chapter, you will learn how to use the full capabilities of the Text Editor and how to add and control text frames for text and leader objects. You will also learn how to use text field objects to automate text values when needed.

In this chapter, we'll cover the following topics:

- Working smarter in the Text Editor
- Using smarter MTEXT
- Using framed text
- Using superscript and subscript
- Taking advantage of text fields
- Controlling MTEXT justification
- Bonus text commands

By the end of this chapter, you will be able to completely control your annotation objects using all the features of these new and old AutoCAD commands.

Working smarter in the Text Editor

First, we need to learn how to use all the features of the Text Editor, including navigation and editing shortcuts, and how to control the formatting using paragraph styles.

Navigating in the Text Editor

In this exercise, we will learn to easily navigate through the text in the Editor using various keyboard shortcuts:

1. Open the 3-1_Smarter Text Editor.dwg file.

2. Using the In-Canvas View Controls, restore the **Custom Model Views | 1-Navigate Text Editor** named view.

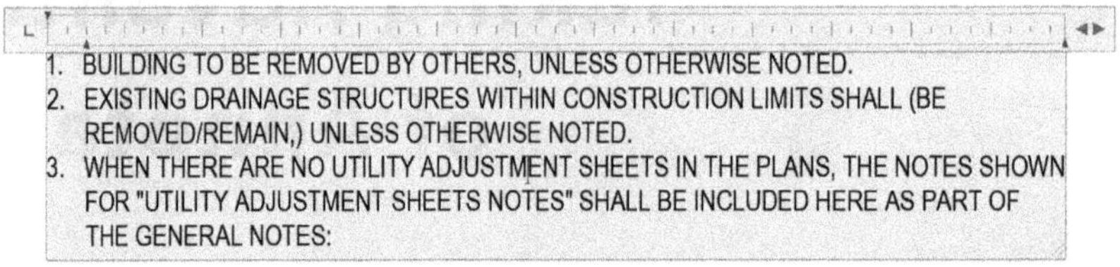

Figure 3.1: Navigation in the Text Editor

3. *Double-left-click* on the existing, numbered TEXT object to open in the Text Editor and allow you to navigate in the editor.

Jump WORD to WORD

1. Use the *Ctrl + ←* or *Ctrl + →* to move the cursor from word to word in each direction using the Text Editor dialog.

Jump PARAGRAPH to PARAGRAPH

1. Use the *Ctrl + ↑* or *Ctrl + ↓* to move the cursor from paragraph to paragraph in each direction using the Text Editor dialog.

2. *Left-click* anywhere in the view window to close the Text Editor.

In the next exercise, we will learn to select various portions of the text object in the Text Editor.

Selecting text in the Text Editor

In this exercise, we will learn how to easily select multiple portions of the text in the Text Editor.

Select a single word

Here we will learn to easily select a single word in the Text Editor.

1. Continue using the 3-1_Smarter Text Editor.dwg file.

2. Using the In-Canvas View Controls, restore the **Custom Model Views | 2-Select a SINGLE WORD** named view.

3. *Double-left-click* on the existing TEXT object, with paragraphs, to open in the Text Editor and allow you to select text in the editor.

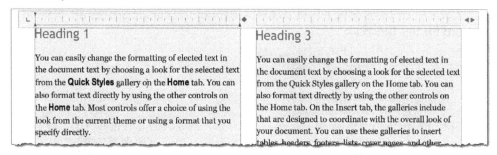

Figure 3.2: Select a single word in the Text Editor

4. *Double-left-click* on a word to select that single word.

Select WORD to WORD

1. Use *Ctrl + Shift + →* or *Ctrl + Shift + ←* to select word-to-word in each direction using the Text Editor dialog.

Select PARAGRAPH to PARAGRAPH

1. Use *Ctrl + Shift + ↑* or *Ctrl + Shift + ↓* to select a paragraph in each direction using the Text Editor dialog.

> **Note**
> You can also *triple-left-click* to select the entire paragraph, similar to other Windows applications.

2. *Left-click* anywhere in the view window to close the Text Editor.

In the next exercise, we will learn to control what portions of the text object are deleted in the Text Editor.

Deleting words in the Text Editor

In this exercise, we will learn to easily control what text is deleted using the Text Editor.

Remove Words to the Right of Cursor

Here we will learn how to remove all words to right of the cursor location in the Text Editor.

1. Continue using the 3-1_Smarter Text Editor.dwg file.

2. Using the In-Canvas View Controls, restore the **Custom Model Views | 2-Select Single Word** named view.

3. *Double-left-click* on the existing TEXT object, with paragraphs, to open the Text Editor, which allows you to delete text using the Editor.

4. Use *Ctrl + Delete* to remove all words to the right of the cursor. Spaces are treated as a word for this function.

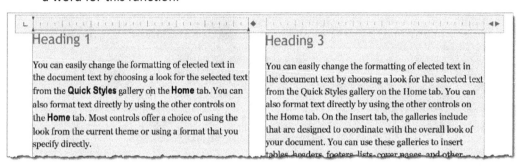

Figure 3.3: Remove words in Text Editor

Remove words to the left of the cursor

1. Use *Ctrl + Backspace* to remove all words to the left of the cursor. Spaces are treated as a word for this function.

2. *Left-click* anywhere in the view window to close the Text Editor.

Control paragraph width

Next, we will learn how to control the paragraph width during the editing process using the Text Editor:

1. Continue using the 3-1_Smarter Text Editor.dwg file.

2. Using the In-Canvas View Controls, restore the **Custom Model Views | 1-Navigate Text Editor** named view.

3. *Double-left-click* on the existing, numbered TEXT object to open the Text Editor and allow you to edit the text in the editor.

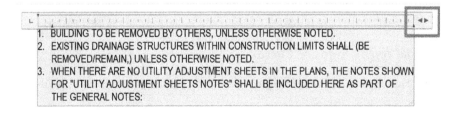

Figure 3.4: Text Editor paragraph width

4. *Double-left-click* on the right margin of the Text Editor ruler to automatically set the paragraph width to the length of the current content.

5. *Left-click* on the Tab icon ⌐ of the Text Editor ruler to modify the Tab style value. The following Tab styles are available:

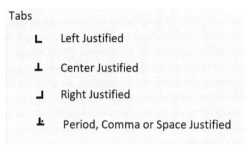

Figure 3.5: Text Editor tab styles

6. You can access the Tab styles using the Tab selection button ⌐ in the Text Editor.

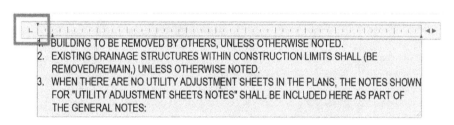

Figure 3.6: Text Editor tab selection button

7. *Left-click* anywhere in the view window to close the Text Editor.

Paragraph styles

Do you want more control over the appearance of your text objects? Have you used all the settings that are available in the Text Editor? This section will teach us how to modify the following "easily overlooked" settings:

1. Continue using the 3-1_Smarter Text Editor.dwg file.

2. Using the In-Canvas View Controls, restore the **Custom Model Views | 2-Select a Single Word** named view.

3. *Double-left-click* on the existing TEXT object, with paragraphs, to open the Text Editor and allow you to edit text in the editor.

4. *Right-click* inside the Text Editor to access the **Paragraph** command and open the **Paragraph** dialog.

5. Review the **Tab**, **Indent**, **Alignment**, and **Line Spacing** settings.

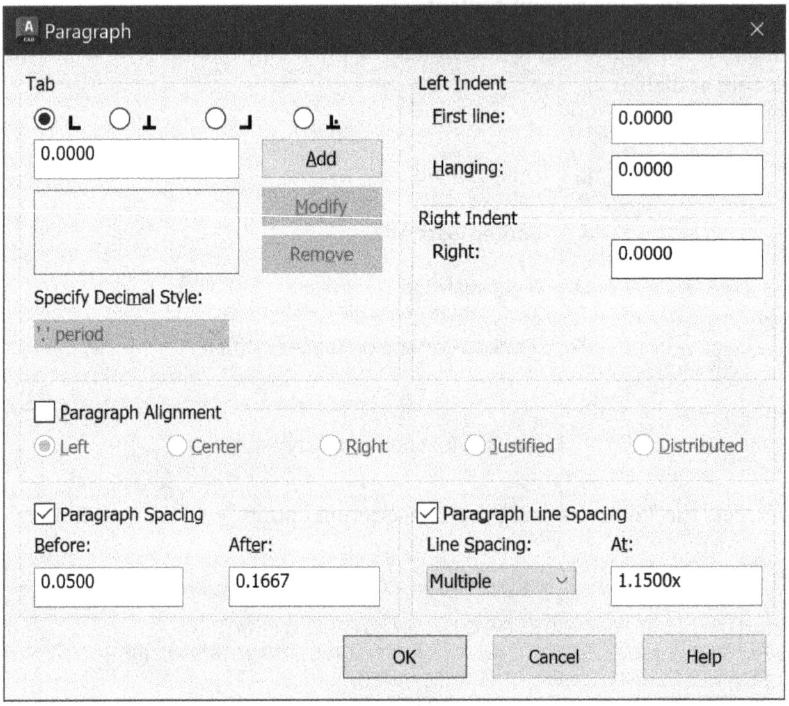

Figure 3.7: The Text Editor Paragraph dialog

6. Click **OK** to close the **Paragraph** dialog.

7. *Left-click* anywhere in the view window to close the Text Editor.

Remove all formatting

You can also select text in the Text Editor and remove all formatting from the selected text:

1. Continue using the `3-1_Smarter Text Editor.dwg` file.

2. *Double-left-click* on the existing TEXT object, with paragraphs, to open the Text Editor and allow you to edit text in the editor.

3. *Left-click* in front of the first paragraph and use *Ctrl + Shift + ↓* to select the first paragraph.

4. *Right-click* inside the Text Editor to access the **Remove Formatting | Remove All Formatting** commands.

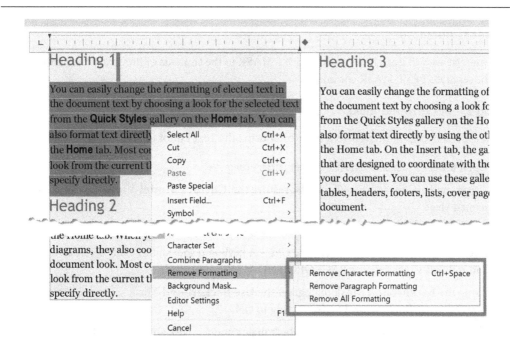

Figure 3.8: Remove all formatting using the Text Editor

5.   *Left-click* anywhere in the view window to close the Text Editor.

> **Note**
> Using this same workflow, you can also access the REMOVE CHARACTER FORMATTING or REMOVE PARAGRAPH FORMATTING commands.

> **Note**
> You can use the *Ctrl + Shift + L* shortcut to change text to lowercase, and the *Ctrl + Shift + U* shortcut to change text to uppercase.

In this section, you learned how to navigate, select, and delete text using the Text Editor and how to control the style and width of your paragraphs. These shortcuts will improve your overall efficiency when editing text objects.

Hiding graphics with text

Are you still creating a wipeout object to create the effect of a "text mask"? If so, a couple of improved methods exist to apply a "text mask." You can use the background mask using the Text Editor or Express Tools | TEXTMASK and TEXTUNMASK.

Using text mask

In this exercise we will learn how to add a TEXT MASK to the contents of the Text Editor, instead of using the separate TEXT MASK command.

Method 1 – background mask

You can apply a background mask to a text object using the Text Editor dialog:

1. Open the 3-2_More Text Commands.dwg file.

2. Using the In-Canvas View Controls, restore the **Custom Model Views | 1-Text Masks** named view.

3. *Double-left-click* on the "FIRST STREET" text object to open the Text Editor dialog.

4. Select all the text in the editor (use the *Ctrl + A* keyboard shortcut to easily select all the text), and *right-click* to access the BACKGROUND MASK command.

5. Using the **Background Mask** dialog, set the **Use Background Mask** setting to ON and the **Use Drawing Background Color** setting to ON.

6. Click **OK** to close the **Background Mask** dialog and *left-click* anywhere in the view window to accept these changes and close the Text Editor dialog.

Figure 3.9: Initial background mask graphics and results

> **Note**
>
> The Border Offset Factor for the background mask is also controlled using the **Background Mask** dialog and can be modified if needed. The default setting is 1.5.

Method 2 – TEXTMASK

The TEXTMASK command hides all objects under a text object by automatically creating a wipeout object to obscure the objects behind the text. The wipeout and text object are created as a group of elements containing the wipeout object and all selected text objects. As a group, if the text is moved, the wipeout will automatically move with it. You can temporarily separate a group of objects using the *Ctrl + Shift + A* keyboard shortcut to toggle **OFF** group recognition. Use *Ctrl + Shift + A* again to toggle **ON** group recognition.

TEXTUNMASK

The TEXTUNMASK command removes the group and wipeout object from the text object.

After editing the text, you can update the mask (WIPEOUT) by using the TEXTMASK command again. Select the text object to be updated and the previous mask object is erased with a new one is created:

1. Continue using the 3-2_More Text Commands.dwg file.

2. Using the In-Canvas View Controls, restore the **Custom Model Views | 1-Text Masks** named view.

3. Using the **Express Tools** ribbon and the expanded **Text** panel, select the **Text Mask** command.

4. Select the "SECOND STREET" text object.

5. The wipeout is added behind the extent of the text object, and a group is created.

TFRAME

Use the TFRAME command to toggle the display of wipeout boundaries:

1. Using the Command Line, key in TFRAME to toggle the display of the wipeout frame for easier editing of the wipeout object.

> **Note**
> TFRAME is also a command option when using the WIPEOUT command.

In the next exercise, we will look at an old "hidden" command that detects TEXT objects with overlapping (or invading) graphics.

TSPACEINVADERS

Use the TSPACEINVADERS command to locate all TEXT objects that have overlapping objects. Overlapping meaning is the same as "invading" objects:

1. Continue using the 3-2_More Text Commands.dwg file.

2. Using the In-Canvas View Controls, restore the **Custom Model Views | 2-TSPACEINVADERS** named view.

3. Using the Command Line, key in the TSPACEINVADERS command.

4. When prompted to select objects, key in ALL and press *Enter* to continue.

5. Key in Y for **Yes** to step through each TEXT object that has overlapping objects. The highlight, but it is difficult to see sometimes, so you might prefer to skip this step and just let the command select all overlapping TEXT objects.

Once the selection set has been defined, you can use any command in AutoCAD to manipulate the selection set. In this example, we will use the TEXT MASK command again:

1. Using the **Express Tools** ribbon and the expanded **Text** panel, select the **Text Mask** command.

2. When prompted to select objects, key in P for **Previous** to use the selection set defined using the TSPACEINVADERS command.

3. All three TEXT objects are masked.

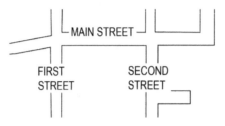

Figure 3.10: TSPACEINVADERS results

> **Note**
>
> TSPACEINVADERS can modify the visibility parameters of some dynamic blocks. Autodesk recommends you save a copy of your file before using it, just in case it affects your dynamic blocks.

In this section, you learned how to apply and control text masks to text objects in the drawing.

In the next section, you will learn how to control the appearance of your text objects using columns and line spacing.

Smarter MTEXT

Next, we will learn how to control the appearance of our MTEXT objects using various settings, such as dynamic columns and line spacing.

Dynamic Columns

One of the best features of the Text Editor is the flexibility in controlling multiple columns in our paragraphs and large paragraph MTEXT objects. Not only can you control the number of columns, but you can also control the spacing. This improves the overall appearance of large amounts of notes needed in a drawing file, and the pagination is dynamic.

1. Continue using the 3-1_Smarter Text Editor.dwg file.

2. Using the In-Canvas View Controls, restore the **Custom Model Views | 2-Select a Single Word** named view.

3. *Double-left-click* on the existing TEXT object, with paragraphs, to open the Text Editor and allow you to edit text in the editor.

4. *Right-click* in the MTEXT dialog to access the **Columns | Dynamic Columns | Auto height** settings.

5. *Left-click* anywhere in the view window to accept this change.

6. *Left-click* anywhere in the view window to close the Text Editor.

Using Grips to edit columns

Many users think it is easier to edit the text and column sizes using GRIPS, and yes, you can control the column sizing using GRIPS, as shown in these steps:

1. Continue using the `3-1_Smarter Text Editor.dwg` file.

2. Select the MTEXT object, with paragraphs, and use the **Column Height** icon ▼ to dynamically modify the column.

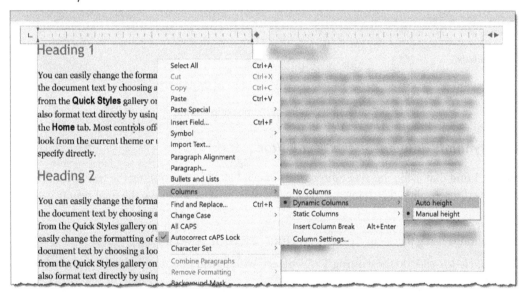

Figure 3.11: Using Dynamic Columns in the Text Editor

3. Use the **Column Width** icon ▶ to modify the width of the columns.

Heading 1

Column Width ▶

Heading 3

You can easily change the formatting of elected text in the document text by choosing a look for the selected text from the **Quick Styles** gallery on the **Home** tab. You can also format text directly by using the other controls on the **Home** tab. Most controls offer a choice of using the look from the current theme or using a format that you specify directly.

You can easily change the formatting of elected text in the document text by choosing a look for the selected text from the Quick Styles gallery on the Home tab. You can also format text directly by using the other controls on the Home tab. On the Insert tab, the galleries include that are designed to coordinate with the overall look of your document. You can use these galleries to insert

from the Quick Styles gallery on the Home tab. You can easily change the formatting of selected text in the document text by choosing a look for the selected text from the Quick Styles gallery on the Home tab. You can also format text directly by using the other controls on the Home tab. When you create pictures, charts, or diagrams, they also coordinate with your current document look. Most controls offer a choice of using the look from the current theme or using a format that you specify directly.

document. You can use these galleries to insert tables, headers, footers, lists, cover pages, and other document building blocks.

▼ **Column Height**

▼ **Column Height**

Figure 3.12: Dynamic columns grip editing

Using Dynamic Columns in the Text Editor

However, I find that using the Text Editor works much better for controlling the text and column sizes. The editor displays the changes dynamically, making it easier to see the final results before accepting the edit:

1. Continue using the `3-1_Smarter Text Editor.dwg` file.
2. *Double-left-click* on the existing TEXT object, with paragraphs, to open the Text Editor and allow you to edit text in the editor.
3. Use the **Width** icon ◆ to control the column heights.
4. Use the **Total Width** icon ◀▶ to control the distance between the dynamic columns.
5. *Left-click* anywhere in the view window to accept these changes.

Smarter text spacing

Do you want to control how the line spacing is handled for MTEXT objects when you use special characters that can be "taller" than regular characters? You can use the system variable TSPACETYPE to automatically control the space between lines of MTEXT.

TSPACETYPE	
Controls the type of line spacing used in multiline text.	
Type: Integer	
Saved in: Registry	
1 (default)	At Least. Adjusts line spacing based on the tallest characters in a line.
2	Exactly. Uses the specified line spacing, regardless of individual character sizes.

1. Continue using the 3-2_More Text Commands.dwg file.

2. Using the In-Canvas View Controls, restore the **Custom Model Views | 3-Control Text Spacing** named view.

3. Using the Command Line, key in TSPACETYPE and change the current setting to 2.

4. Using the **Home** ribbon and the **Annotation** panel, select the **Multiline Text** command A and key in a new paragraph of text.

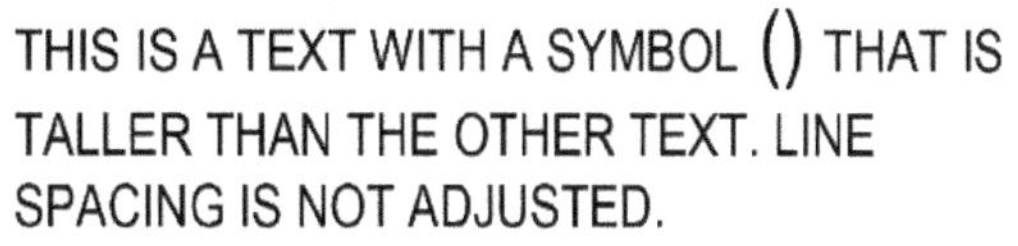

THIS IS A TEXT WITH A SYMBOL () THAT IS
TALLER THAN THE OTHER TEXT. LINE
SPACING IS NOT ADJUSTED.

Figure 3.13: New text content

5. *Left-click* anywhere in the view window to complete the MTEXT command.

Here, you can see the differences between the line spacing.

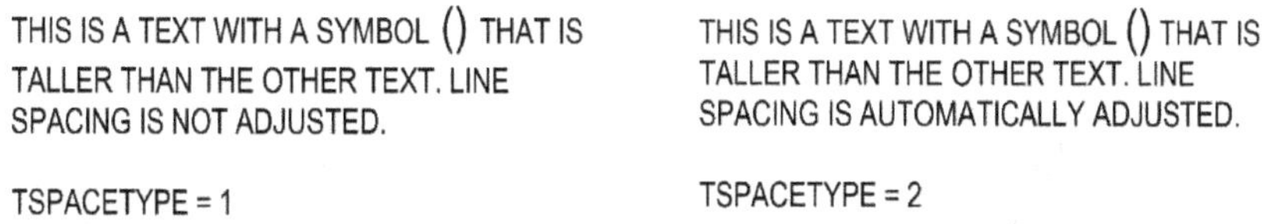

THIS IS A TEXT WITH A SYMBOL () THAT IS
TALLER THAN THE OTHER TEXT. LINE
SPACING IS NOT ADJUSTED.

TSPACETYPE = 1

THIS IS A TEXT WITH A SYMBOL () THAT IS
TALLER THAN THE OTHER TEXT. LINE
SPACING IS AUTOMATICALLY ADJUSTED.

TSPACETYPE = 2

Figure 3.14: TSPACETYPE results

In this section, you learned how to control a paragraph's number of columns and the line spacing.

The next section will teach us how to place smarter text frames around your objects.

Using framed text

Your text and leader objects often require a frame to make them stand out in a drawing. It is important that these text frames update automatically when the text is edited. Continue reading to learn how to make your text frames auto-update as needed.

Smart text frames

First, let's examine how you can use text shapes that automatically update when the text is edited.

MTEXT

Do you ever need a box shape around your text that is smart enough to grow when the text is edited? I have seen many suggestions that tell you to use a table with one column and one row, but I think I have a better solution:

1. Continue using the 3-2_More Text Commands.dwg file.

2. Using the In-Canvas View Controls, restore the **Custom Model Views | 4-Framed Text** named view.

3. Select the single-line MTEXT object.

4. *Right-click* to access the **Properties** command from the pop-up menu.

5. Using the **Properties** dialog, scroll to the **Text** panel, and locate the **Text Frame** setting.

6. Toggle this setting to **Yes** to add a text frame to this MTEXT object.

> **Note**
>
> The offset distance of the text frame is determined by the **Landing Gap** setting using the **Properties** dialog.

Next, let's look at another method that allows you to define a standard style to create framed MTEXT objects using a LEADER object.

Leader frame styles

In this exercise, we will use a leader style to create a leader with no leader line. Yes, it is possible:

1. Continue using the 3-2_More Text Commands.dwg file.

2. Using the **Annotate** ribbon and the **Leaders** panel, select the dialog launcher ⬎ to open the **MultiLeader Style Manager** dialog.

3. Select the **New...** button to create a new style named **Standard-BOXED**.

4. Verify that the **Annotative** setting is turned on and click the **Continue** button.

5. Using the **Modify MultiLeader Style** dialog, select the **Leader Format** tab and set **Type** to **None**.

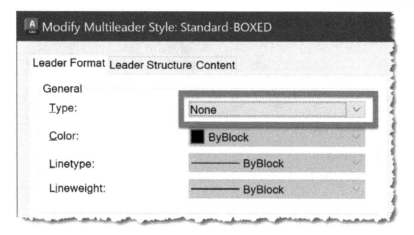

Figure 3.15: Leader format | type

6. Using the **Leader Structure** tab, clear all settings for **Constraints** and **Landing settings**, and activate the **Annotative** setting.

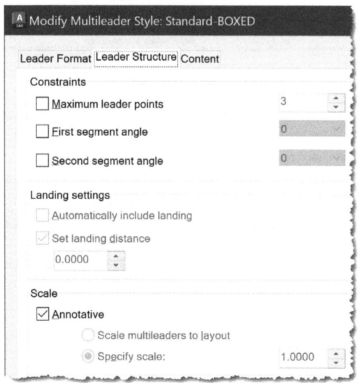

Figure 3.16: Leader structure | constraints

7. Using the **Content** tab, turn **Frame Text** on.

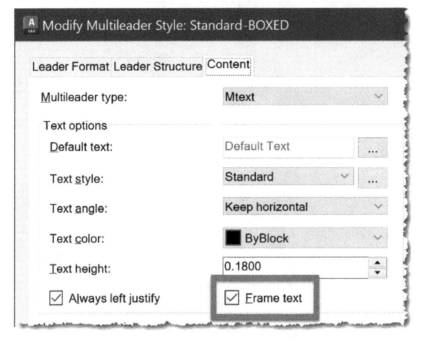

Figure 3.17: Leader Content Settings

8. Click **OK** to save these changes and click the **Set Current** button to make the new **Standard-BOXED** STYLE the active leader style.

9. Click the **Close** button to close the dialog.

10. Using the **Annotate** ribbon and the **Leaders** panel, select the **MultiLeader** command. ⌒°

11. Use the *left-click* in the view to define the area for the LEADER object text.

12. Using the Text Editor, key in the following text: THIS IS A FRAMED TEXT OBJECT USING MLEADER.

13. *Left-click* in the view window to complete the command and close the Text Editor.

14. Review the TEXT FRAME object, which is actually a LEADER object without the linear and arrow graphics.

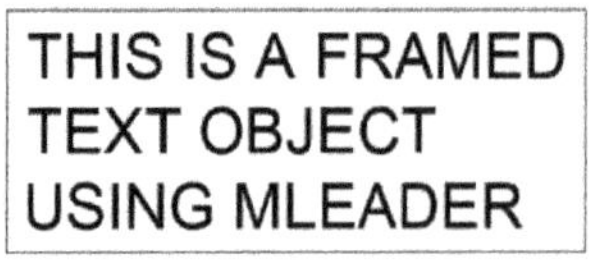

THIS IS A FRAMED
TEXT OBJECT
USING MLEADER

Figure 3.18: TEXT FRAME using the LEADER object

In the next section, we will learn how to apply the superscript and subscript text appearance to various text examples.

Using superscript and subscript

Next, we will learn how to use the stacked fractions to quickly and easily create a superscript or subscript text appearance without using the Text Editor dialog.

These examples are using the following **Stack Properties**:

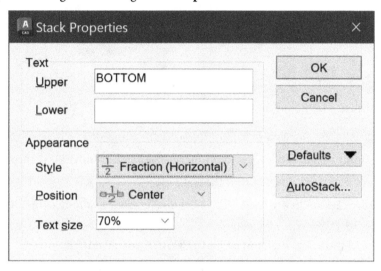

Figure 3.19: Stack fraction properties

Quick superscript text

For the first example, we will use the SUPERSCRIPT | S``TACK option and apply it to numbered text.

1. Continue using the 3-2_More Text Commands.dwg file.

2. Using the In-Canvas View Controls, restore the **Custom Model Views | 5-Superscript Text** named view.

3. Using the **Home** ribbon and the **Annotation** panel, select the **Multiline Text** command. A

4. *Left-click* in the view window to define the text area.

5. Using the Text Editor, key in TOPBOTTOM^ and *left-click and drag* to highlight only the BOTTOM^ characters.

6. *Right-click* inside the Text Editor to access the **STACK** command.

7.   *Left-click* anywhere in the view window to complete the command.

$$\text{TOP}^{\text{BOTTOM}}$$

Figure 3.20: Superscript 1

8.   You can also highlight the entire text string for this result.

TOP
BOTTOM

Figure 3.21: Stack text 1

Subscript text

Next, let's apply the same technique to subscript:

1.   Continue using the 3-2_More Text Commands.dwg file.
2.   Using the **Home** ribbon and the **Annotation** panel, select the **Multiline Text** command. **A**
3.   *Left-click* in the view window to define the text area.
4.   Using the Text Editor, key in TOP^BOTTOM and *left-click and drag* to highlight only the ^BOTTOM characters.
5.   *Right-click* inside the Text Editor to access the **STACK** command.
6.   *Left-click* anywhere in the view window to complete the command.

$$\text{TOP}_{\text{BOTTOM}}$$

Figure 3.22: Subscript 1

7.   You can also highlight the entire text string for this result.

More examples

Next, let's apply these same techniques to other text combinations.

Option 1 – using stack fractions

In this exercise, we will apply stacked fractions to text content in the Text Editor.

1.   Continue using the 3-2_More Text Commands.dwg file.
2.   Using the **Home** ribbon and the **Annotation** panel, select the **Multiline Text** command. **A**

3.   *Left-click* in the view window to define the text area.

4.   Using the Text Editor, key in TOP/BOTTOM and *left-click and drag* to highlight only the /BOTTOM characters.

5.   *Right-click* inside the Text Editor to access the **STACK** command.

6.   *Left-click* anywhere in the view window to complete the command.

Figure 3.23: Subscript 2

7.   You can also highlight the entire text string to stack any text.

Figure 3.24: Stack text 3

8.   You can modify the **Stack Properties** settings to use the **Fraction (Diagonal)** style for this result.

Figure 3.25: Stack text 4

Option 2 – forcing stacked fractions

In this exercise, we will use the "#" character to force stacked fractions to text content in the Text Editor.

1.   Continue using the 3-2_More Text Commands.dwg file.

2.   Using the **Home** ribbon and the **Annotation** panel, select the **Multiline Text** command. A

3.   *Left-click* in the view window to define the text area.

4.   Using the Text Editor, key in TOP#BOTTOM and *left-click and drag* to highlight only the #BOTTOM characters.

5.   *Right-click* inside the Text Editor to access the **STACK** command.

6.   *Left-click* anywhere in the view window to complete the command.

Figure 3.26: Subscript 3

In the next section, we will learn how to use TEXT FIELD objects for smarter objects in our drawings.

Controlling MTEXT center justification

One of the best reasons for using MTEXT is the ability to manually adjust the text width of the paragraph by dragging a grip and resizing or stretching the paragraph width. This has changed slightly if the text is center justified. Stretching center justified MTEXT using grips moves the center origin of the entire MTEXT object.

To re-size MTEXT, AutoCAD expects the user to use the ruler in the in-place Text Editor, not by grips. Sorry, but I agree that it takes too many steps, and no user will choose this method as the first option.

To revert the MTEXT grip editing functionality to the "old way," you need to change the CENTERMT system variable. Follow these steps:

CENTERMT	
Controls how grips stretch MTEXT that has a centered justification. CENTERMT does not apply to stretching multiline text by using the ruler in the In-Place Text Editor.	
Type: Integer	
Saved in: Registry	
0 (default)	When stretching the MTEXT with grips, the center grip also moves in the same direction.
1	When stretching the MTEXT with grips, the center grip stays in place while the text is re-sized.

1. Continue using the 3-2_More TEXT Commands.dwg file.

2. Using the In-Canvas View Controls, restore the **Custom Model Views | 6-Center Text** named view.

3. Select the MTEXT object located in the center of the TOP rectangle shape.

4. Using grips, select any of the corner grips and stretch the text paragraph width smaller.

   Notice how the center insertion point of the text is moved by default.

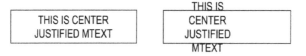

Figure 3.27: CENTERMT = 0

5. Using the Command Line, key in CENTERMT and change the value to 1.

6. Select the other MTEXT object located in the center of the rectangle shape.

7. Using grips, select any of the corner grips and stretch the text paragraph width smaller.

   Notice how the center insertion point of the text remains in the center.

Figure 3.28: CENTERMT = 1

In this exercise, you learned how to better control the text location when changing a text object's justification.

In the next section, we will learn how to control text, dimensions, and leader objects with some of the older command options that are easily overlooked.

Using smart fields

For the first example, we will use a TEXT FIELD object to populate shapes with automatic areas that will update when the selected object is modified.

INSERT FIELD	Command Locations
Ribbon	Insert | Data | Field
Text Editor dialog	*Right-click* | Insert Field
Command Line	FIELD (FIE)

Automatic areas using fields

If you need to label areas in your drawing, you can use TEXT FIELD objects along with closed object types to "automatically" label the area and keep the area value up to date:

1. Open the 3-3_Using Fields.dwg file.

2. Using the In-Canvas View Controls, restore the **Custom Model Views | 1-Object Area** named view.

3. Using the **Home** ribbon and the **Annotation** panel, select the **MTEXT** command.

4. Using the Text Editor, key in AREA = and *right-click* to access the **INSERT FIELD** command.

5. Using the **Field** dialog, modify the following settings:

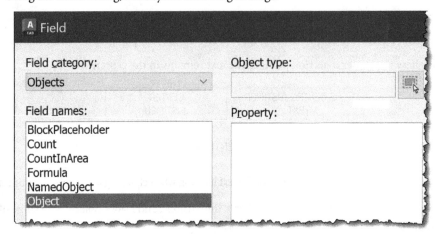

Figure 3.29: FIELD dialog

6. Click the **Object** button 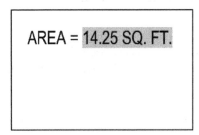 and select the OBJECT 1 closed polyline. This will define the **Object** type to a POLYLINE.

7. Using the **Field** dialog, select **Property | Area** and change **Format** to **Architectural**. Also, modify **Precision** to 0.00.

8. Using the **Preview** field, verify that the field value displayed includes the SQ. FT. suffix and click **OK** to close the dialog.

9. *Left-click* anywhere in the view window to complete the command.

AREA = 14.25 SQ. FT.

Figure 3.30: TEXT FIELD results

TEXT FIELD objects display with a "gray" background that will not plot. This background display makes it easier to distinguish between normal TEXT and TEXT FIELDS.

You can turn off this "gray" background display if needed; however, this makes it impossible to distinguish between the two object types.

TEXT FIELD options

There are a couple of personal options available that each user can customize to control how their TEXT FIELDS function:

1. Continue using the `3-3_Using Fields.dwg` file.

2. Place your cursor in the view window and *right-click* to access the OPTIONS command.

Figure 3.31: Field user preferences

3. Using the **Options** dialog, select the **User Preferences** tab. In the lower-left corner, find the **Field Options** section.

4. Turn OFF the **Display background of fields** setting if you want to remove the "gray" background in the TEXT FIELD. Again, I recommend leaving this ON for most situations.

5. Select the **Field Update Settings** button to control when your TEXT FIELD objects will update. If you use a lot of TEXT FIELDS, performance may be an issue, and you can improve your performance by controlling when all the TEXT FIELDS update using these settings.

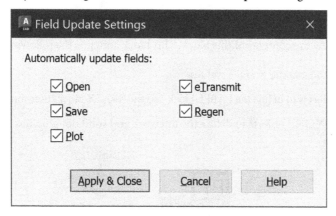

Figure 3.32: Field Update Settings

6. Click **Apply & Close** to close the **Field Update Settings** dialog, and click **OK** to close the **Options** dialog.

> **Note**
>
> You can also use this example to place the TEXT FIELD in a multileader object type.

> **Note**
>
> You can add a TEXT FIELD to an old DTEXT object using the shortcut *Ctrl + F* to insert a FIELD.

In the next example, we will create a block that contains automatic coordinate labels using TEXT FIELD objects.

Automatic coordinate labels

Many users have asked me how to label a drawing coordinate automatically in AutoCAD. Have you figured out a good way to do this yet?

You can use an ordinate dimension, but there is a better way if you make a "smart" but dynamic block to label points in your drawing.

In this exercise, we will learn how to use the TEXT FIELD object in a block to automatically label file coordinates:

1. Continue using the `3-3_Using Fields.dwg` file.

2. Using the In-Canvas View Controls, restore the **Custom Model Views | 2- XY Coordinates** named view.

3. Select the XY block and *right-click* to access the **Block Editor** command.

First, we need to review the contents of this block. This block contains the following:

- TEXT objects to label the X and Y values

- ATTRIBUTE objects to define the LABEL block and the X and Y block insertion coordinate values

- LINE and POLYLINE objects to define the linework and solid dot graphics

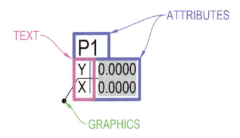

Figure 3.33: Coordinate block objects

Let's review how these objects were created. First, we will place the first attribute object for the block label:

1. Using the **Home** ribbon and the expanded **Block** panel, select the **Define Attributes** command.

2. Using the **Attribute Definition** dialog, define the following values:

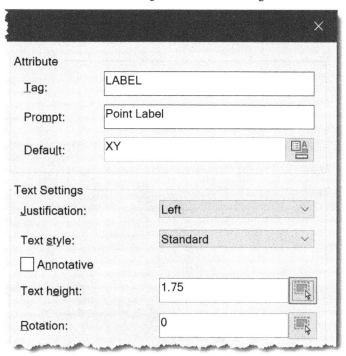

Figure 3.34: Point label attribute definition values

3. Click **OK** to close the dialog and save these changes.

4. *Left-click* to place the label attribute as needed.

Since the attributes for this block are already provided, you can UNDO (*Ctrl + Z*) this placement and keep the original or make a new block with your attributes.

Next, we need to create the attributes that will use a TEXT FIELD to populate with the X and Y values:

1. Using the **Home** ribbon and the expanded **Block** panel, select the **Define Attributes** command.

2. Using the **Attribute Definition** dialog, define the following values:

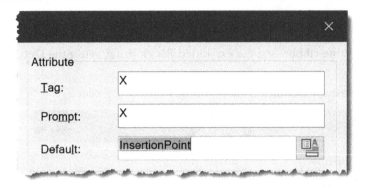

Figure 3.35: X Attribute definition values

3.   Use the **Insert Field** button 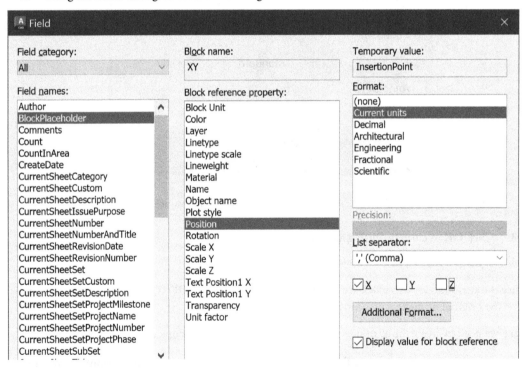 to add a text field object to the **Default:** setting in this dialog.

4.   Using the **Field** dialog, select the following:

Figure 3.36: The BlockPlaceHolder field values

5. Continue using the **Field** dialog to set the following settings:

Text Settings

Justification:	Left
Text style:	Standard
☐ Annotative	
Text height:	1.75
Rotation:	0
Boundary width:	0.0000

Figure 3.37: Attribute definition text settings

6. Click **OK** to close the dialog and save these changes.

7. Repeat *Steps 7-12* to add the Y attribute object if needed.

Next, we need to add a POINT parameter and a STRETCH action to this block so we can move the text but leave the DOT insertion point as-is since this is the point that displays the coordinate value:

1. Using the **Block Editor** and the **Block Authoring** palette, select the **Parameters** tab.

2. Select the **POINT** parameter and OSNAP to the **Intersection** of the 4 lines at **P1**.

3. Using the **Block Authoring** palette, select the **Action** tab and select the **STRETCH** action.

4. Select the previous **POINT** parameter to apply the **STRETCH** action.

Next, we must define the STRETCH boundary to control what objects move or stretch with the text objects. Define the STRETCH boundary as shown here.

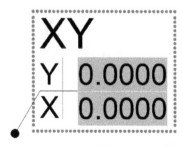

Figure 3.38: The STRETCH action boundary

We will use the STRETCH action rather than the MOVE action because we need the diagonal line to STRETCH, not MOVE, when the TEXT is moved. As long as the TEXT objects are inside the STRETCH action boundary, they will move during the STRETCH action.

Now, we need to select the objects that will move (stretch) with the TEXT objects. Select the following objects to move.

Figure 3.39: The STRETCH action objects

I recommend renaming your dynamic block definitions with logical names to help with future editing. First, we will rename the POINT parameter POSITION1:

1. Select the **POINT** parameter and, using the **Properties** dialog, scroll to the **Property Labels** section and modify **Position Name | Text Position1**.

    Next, we need to rename the **STRETCH** action.

2. *Right-click* on the **STRETCH** action icon to access the **Rename Action** command and modify the name **STRETCH1** to Text Stretch1.

3. Move the new **STRETCH** icon  to be near the previous POINT parameter so it is more obvious that they are related, as shown here. *Left-click and drag* on the icon to move it to the new location.

Figure 3.40: STRETCH action icon

4. Using the **Block Editor** ribbon and the **Close** panel, select the **Close Block Editor** command and select **Save the Changes**.

Now, we are ready to place and test our new block:

1. Using the **Home** ribbon, and the **Block** panel, select the **Insert** command *drop-down-list*, and select the **XY** block.

2. *Left-click* in a blank area of the drawing to place the new block.

3. Using the **Edit Attributes** dialog, key in `POINT 1` for the **Point Label** field and click **OK** to close the dialog.

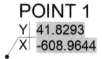

Figure 3.41: XY Coordinate Block results

The block is placed, and the POINT 1 label and the coordinate values are automatically populated in the block. The coordinate values reflect the XY coordinate of the block insertion point.

Next, we need to test our dynamic block parameters:

1. Select the new block and, using grips, select the **Text** grip and move the TEXT object to another location. Notice the **DOT** insertion point does not move.

2. Select the **DOT** insertion point grip and move the **Block** insertion point. Notice that the XY fields do not update automatically.

   Remember, TEXT FIELDS update using the following settings by default:

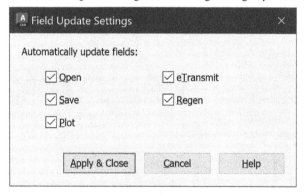

Figure 3.42: Field update settings

3. Key in the REGEN command to update the field values.

> **Note**
>
> I have provided an XYZ block with this file for your review using the block XYZ.

FIELDEVAL	
The setting is stored as a bit code using the sum of the following values:	
Type: Integer	
Saved in: Drawing	
Initial Value: 31	
0	Not updated
1	Updated on open
2	Updated on save
4	Updated on plot
8	Updated on use of ETRANSMIT
16	Updated on regeneration

In this section, you learned how to use smarter text using TEXT FIELDS to create text values to automate graphic-related text objects.

In the next section, you will learn how to control text justification to prevent text object movement.

Bonus commands

Next, let's examine some commands and settings that can affect your text, dimensions, and leader objects.

TEXTTOFRONT

The first command I want to demonstrate is the TEXTTOFRONT command, which is available as a key in or by using the somewhat hidden Draw Order commands. I personally wish this command was more accessible, so you might consider how you can customize your interface to bring this command to a *right-click* menu or a custom ribbon or toolbar.

TEXTTOFRONT	Command Locations
Ribbon	Home | Modify | Draw Order | Bring All Annotations to Front
Command Line	TEXTTOFRONT (TEXTT)

1. Open the 3-4_Bonus TEXT Commands.dwg file.

2. Using the In-Canvas View Controls, restore the **Custom Model Views | 1-TEXTTOFRONT** named view.

Here, you can see that I have several solid-filled shapes with text, dimensions, and leaders buried under the solid graphics. Let's look at this a little closer.

3. Using the Status Bar, select the "hamburger" icon ≡ in the lower right corner and turn ON the **Transparency** setting.

4. Turn ON the **Transparency** display and notice the annotation objects beneath the solid-filled objects. Then, turn OFF the **Transparency** display for the next steps.

5. Using the **Home** ribbon, and the expanded **Modify** panel, select the **DrawOrder** command *drop-down-list*, and select the **Bring Text To Front** command.

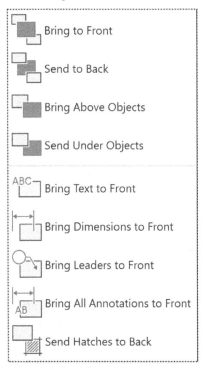

Figure 3.43: The DrawOrder commands

6. Use the ENTER key to bring all annotation objects to the top of all other object types.

Now, you can see why I usually key in the TEXTTOFRONT command using the Command Line. This command is too hard to access! You can see all the annotation objects on top of the solid fills.

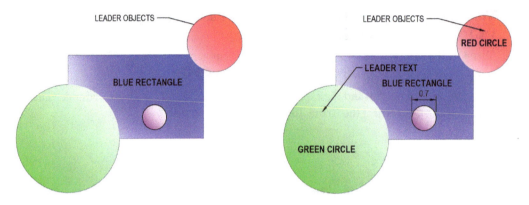

Figure 3.44: TEXTTOFRONT results

> **Note**
>
> The command options for **Text, Dimensions, Leaders**, and **All** can be used to control the individual object types as needed.

The next command I want to demonstrate is the TORIENT command, which is also buried in the Express Tools.

> **Note**
>
> If you do not have an Express Tools ribbon, re-visit the installation of your AutoCAD and add the Express Tools installation. These are some of the most efficient and productive commands in AutoCAD.

Rotate text (TORIENT)

We can use the EXPRESS TOOLS | ROTATE TEXT (TORIENT) command to make unreadable text more readable.

TORIENT	Command Locations
Ribbon	Express Tools | Text | Modify Text | Rotate Text
Command Line	TORIENT (TORI)

In this exercise, we will change the orientation of all TEXT objects that are not right-reading.

1. Continue using the 3-4_Bonus TEXT Commands.dwg file.

2. Using the In-Canvas View Controls, restore the **Custom Model Views | 2-More Readable Text** named view.

    Notice that some of the text is not readable using this view direction. We have backward and upside-down text.

Figure 3.45: TORIENT objects

3. Using the **Express Tools** ribbon, the **Text** panel, and the **Modify Text** command *drop-down-list*, access the **Rotate Text** (TORIENT) command.

4. Using the Command Line, key in ALL to evaluate all text objects in this file and use the *Enter* key to complete the command.

Note the changes in the text orientation to correct any non-readable text from the bottom of the right direction.

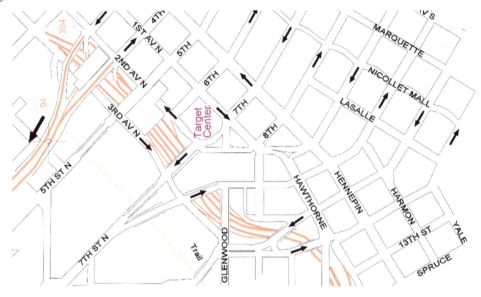

Figure 3.46: TORIENT command results

The next command we will learn to use is the TJUST command, which modifies the justification of a text object without changing the text location during the modification.

Justify Text (TJUST)

Have you ever changed the text justification of a text object only to have the text location change? Of course, we all have.

You can use the **Express Tools | Justify Text** (TJUST) command to avoid these unwanted results.

TJUST	Command Locations
Ribbon	Express Tools	Text	Modify Text	Justify Text
Command Line	TJUST (TJ)

1. Continue using the 3-4_Bonus TEXT Commands.dwg file.

2. Using the In-Canvas View Controls, restore the **Custom Model Views | 3-Justify Text** named view.

   Notice that all the area text objects are CENTERED in the rectangle shapes. These text objects are not MIDDLE-CENTER justified, but TOP-LEFT justified, and we want to change them to MIDDLE-CENTER so we can edit them without losing the CENTER location at the CENTER of the shapes.

3. *Double left-click* on one of the text objects and using the Text Editor, modify the text to the following.

    `TOP AREA = 833.33 SQ. FT.`

    Notice this text object is no longer centered. We will modify the justification to be MIDDLE-CENTER justified.

4. Select the edited text object and *right-click* to access the **Select Similar** command.

5. Using the **Properties** dialog, locate the Text section and change the **Justify** setting to **Middle Center** for all three text objects. Notice how the text location changes.

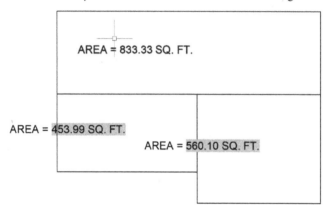

Figure 3.47: TJUST properties results

6. Use the **UNDO** command (*Ctrl + Z*) to select the justification to its original setting and select the text locations to the center of the shapes.

7. Using the **Express Tools** ribbon, locate the **Text** panel, and use the **Modify Text** *drop-down-list* to access the **Justify Text** (`TJUST`) command.

8. Select all three text objects and key in MC to change the justification to **Middle Center**.

9. Select one of the text objects and note that the justification is now set to **Middle Center,** and the text location did not change.

> **Note**
>
> You can control what properties are compared when using the SELECT SIMILAR command using the system variable SELECTSIMILARMODE.

In the next command, we will learn how to use the TEXTALIGN command to easily align text objects without struggling through the normal ACAD Align command.

Text Align (TEXTALIGN)

The TEXT ALIGN command is often overlooked in the Annotate ribbon. So, let's look at how we can use this command instead of the normal object Align command.

TEXTALIGN	Command Locations
Ribbon	Annotate | Text | Text Align
Command Line	TEXTALIGN (TEXTA)

In this exercise, we will use the TEXTALIGN command to easily align TEXT objects, which is much easier than using the ALIGN command.

1. Continue using the 3-4_Bonus TEXT Commands.dwg file.
2. Using the In-Canvas View Controls, restore the **Custom Model Views | 4-Align Text** named view.

Horizontal alignment

Notice that the text objects are not aligned, and we want to align them horizontally. Use the following steps to change their alignment to horizontal:

Figure 3.48: Initial TEXT ALIGN objects

1. Using the **Annotate** ribbon, locate the **Text** panel and select the **Text Align** command.
2. Key in O for the **Options** setting, and key in H to set the alignment mode to **Horizontal**.
3. Select all the ROOM NUMBER text objects and select ROOM 101 as the text object to align the others to.
4. Move the cursor until all ROOM NAME objects are aligned horizontally and *left-click* to accept and complete the command.

ROOM 101   ROOM 102   ROOM 103   ROOM 104   ROOM 105   ROOM 106

Figure 3.49: TEXT ALIGN horizontal results

Vertical alignment

Notice that the text objects are not aligned. We want them to be aligned vertically. Use the following steps to change their alignment to vertical:

1. Continue using the 3-4_Bonus TEXT Commands.dwg file.

2. Using the In-Canvas View Controls, restore the **Custom Model Views | 5-Vertical Align** named view.

1.   WHERE APPLICABLE DIMENSIONS SHOWN ON LONG SECTIONS ARE PLAN LENGTHS.

    2.   ALL GASKETS ARE 3.2mm THICK (COMPRESSED) UNLESS NOTED OTHERWISE.
  3.   ALL FITTING TO FITTING DIMENSIONS INCLUDE 3mm WELD GAPS.
4.   CONTOURS REFLECT EXISTING GROUND LEVELS.
     5.   INTERSECTION POINTS (IP's) SHALL BE PEGGED OUT AND VERIFIED.
  6.   MAXIMUM MISALIGNMENT OF ANY SUPPORT FROM A STRAIGHT LINE BETWEEN I.P.'s HORIZONTAL.
    7.   INSULATION SHALL BE IN ACCORDANCE WITH THE FOLLOWING SPECIFICATION AND DRAWINGS:
8.   CONSTANT WEIGHT SUPPORT TO BE INSTALLED HORIZONTAL IN THE COLD POSITION.

    9.   CHANGES IN DIRECTION OR SLOPE TO BE FULL OR TRIMMED ELBOWS UNLESS SINGLE OR DOUBLE MITER BENDS ARE SPECIFIED
 10.   CONTRACTOR SHALL SUBMIT CONSTRUCTION PLAN AND METHODOLOGY FOR EACH AREA.

   11.   CONTRACTOR SHALL SUBMIT HYDROTEST PLAN AND METHODOLOGY FOR EACH AREA.

Figure 3.50: Initial TEXT ALIGN Objects

3. Using the **Annotate** ribbon, locate the **Text** panel and select the **Text Align** command.

4. Key in O for the **Options** setting, and key in V to set the alignment mode to **Vertical**.

5. Select all the NOTE text objects and select NOTE #1 as the text object to align the others to.

6. Move the cursor until all NOTE text objects are aligned vertically and *left-click* to accept and complete the command.

1.   WHERE APPLICABLE DIMENSIONS SHOWN ON LONG SECTIONS ARE PLAN LENGTHS.

2.   ALL GASKETS ARE 3.2mm THICK (COMPRESSED) UNLESS NOTED OTHERWISE.
3.   ALL FITTING TO FITTING DIMENSIONS INCLUDE 3mm WELD GAPS.
4.   CONTOURS REFLECT EXISTING GROUND LEVELS.
5.   INTERSECTION POINTS (IP's) SHALL BE PEGGED OUT AND VERIFIED.
6.   MAXIMUM MISALIGNMENT OF ANY SUPPORT FROM A STRAIGHT LINE BETWEEN I.P.'s HORIZONTAL.
7.   INSULATION SHALL BE IN ACCORDANCE WITH THE FOLLOWING SPECIFICATION AND DRAWINGS:
8.   CONSTANT WEIGHT SUPPORT TO BE INSTALLED HORIZONTAL IN THE COLD POSITION.

9.   CHANGES IN DIRECTION OR SLOPE TO BE FULL OR TRIMMED ELBOWS UNLESS SINGLE OR DOUBLE MITER BENDS ARE SPECIFIED
10.   CONTRACTOR SHALL SUBMIT CONSTRUCTION PLAN AND METHODOLOGY FOR EACH AREA.

11.   CONTRACTOR SHALL SUBMIT HYDROTEST PLAN AND METHODOLOGY FOR EACH AREA.

Figure 3.51: TEXT ALIGN vertical results

Next, let's control the spacing of these vertically aligned text objects.

Current vertical

We can use the TEXTALIGN command to align and re-distribute the text objects so the line spacing is equal:

1. Continue using the `3-4_Bonus TEXT Commands.dwg` file.

2. Undo *(Ctrl + Z)* the previous vertical alignment modification.

3. Using the **Annotate** ribbon, locate the **Text** panel and select the **Text Align** command.

4. Key in O for the **Options** setting, and key in D to set the alignment mode to **Distribute**.

5. Select NOTE #1 as the reference object to align all NOTE text objects.

6. Move the cursor down until all NOTE text objects are aligned and spaced evenly in a vertical direction.

7. *Left-click* to accept and complete the command.

1.	WHERE APPLICABLE DIMENSIONS SHOWN ON LONG SECTIONS ARE PLAN LENGTHS.
2.	ALL GASKETS ARE 3.2mm THICK (COMPRESSED) UNLESS NOTED OTHERWISE.
3.	ALL FITTING TO FITTING DIMENSIONS INCLUDE 3mm WELD GAPS.
4.	CONTOURS REFLECT EXISTING GROUND LEVELS.
5.	INTERSECTION POINTS (IP's) SHALL BE PEGGED OUT AND VERIFIED.
6.	MAXIMUM MISALIGNMENT OF ANY SUPPORT FROM A STRAIGHT LINE BETWEEN I.P.'s HORIZONTAL.
7.	INSULATION SHALL BE IN ACCORDANCE WITH THE FOLLOWING SPECIFICATION AND DRAWINGS:
8.	CONSTANT WEIGHT SUPPORT TO BE INSTALLED HORIZONTAL IN THE COLD POSITION.
9.	CHANGES IN DIRECTION OR SLOPE TO BE FULL OR TRIMMED ELBOWS UNLESS SINGLE OR DOUBLE MITER BENDS ARE SPECIFIED
10.	CONTRACTOR SHALL SUBMIT CONSTRUCTION PLAN AND METHODOLOGY FOR EACH AREA.
11.	CONTRACTOR SHALL SUBMIT HYDROTEST PLAN AND METHODOLOGY FOR EACH AREA.

Figure 3.52: TEXT ALIGN Distribute results

The following system variable can be defined to specify your preferred default action.

TEXTALIGNMODE	
Stores the alignment option for aligned text	
Type: Integer	
Saved in: Registry	
0	Top Left
	Left-justified at the top of the text (horizontally oriented text only).
1	Top Center
	Centered at the top of the text (horizontally oriented text only).
2	Top Right
	Right-justified at the top of the text (horizontally oriented text only).
3	Middle Left
	Left-justified at the middle of the text (horizontally oriented text only).
4	Middle Center
	Centered both horizontally and vertically in the middle of the text (horizontally oriented text only)
5	Middle Right
	Right-justified at the middle of the text (horizontally oriented text only).
6	Bottom Left
	Left-justified at the baseline (horizontally oriented text only)
7	Bottom Center
	Centered on the baseline (horizontally oriented text only).
8	Bottom Right
	Right-justified at the baseline (horizontally oriented text only).
9 (default)	Left
	Left-justified at the baseline.
10	Center
	Aligned from the horizontal center of the baseline.
11	Right
	Right-justified at the baseline.

TEXTLAYER

Use the TEXTLAYER system variable to define the default layer for new TEXT objects in a drawing.

TEXTLAYER	
Defines the default layer for all new TEXT objects.	
Type: String	
Saved in: Drawing	
"." use current (default)	Use the current layer.
layer name	Use the defined *layer name*.
Note: If the defined layer name does not exist, it is created automatically.	

In these exercises, you learned how to control text objects using the Express Tool commands to display, rotate, and align annotation objects.

Summary

In this chapter, we examined how to use the Text Editor more efficiently using navigation shortcuts, paragraph styles, and columns. We also learned to place smarter text objects using text fields for automatic text values. When placing text with frames, we learned to use both text and leader objects to frame individual text objects automatically.

We finished this chapter learning how to use the Express Tool commands to control the display order for annotation objects and how to rotate, justify, and align text objects more easily.

In the next chapter, we will examine how to take advantage of some of the advanced dimensioning features, including leaders, linear dimensions, and angular dimensions, and how to use some of the hidden commands available to control these dimensions. We will also learn how to control the size of our annotation using an annotation scale.

4

Making the Most of Dimensions

In this chapter, you will learn how to place and modify various dimension types and styles to improve efficiency using these tools. You will also learn how to control appearance, standards, and settings when using the dimension tools. Once you have learned how to place dimensions more quickly, you must also learn how to modify these dimension objects to control their appearance and relationships with other objects in the drawing.

In this chapter, we'll cover the following topics:

- Working more efficiently with leaders
- Using the hidden features of dimension commands
- Using the new smart centerlines
- Learning to manage your annotation scales
- Learning to use obscure dimension commands

By the end of this chapter, you will be able to fully use the dimensioning commands and improve your overall understanding of how to control those dimensions.

Leader power commands

In this first section, we will look at some leader commands that will improve your daily use of LEADER objects.

Adding a leader to MTEXT

First, let's learn how to add a leader to existing MTEXT objects to avoid recreating text objects. From AutoCAD 2023 onward, you can add a leader to an existing MTEXT and convert it to a LEADER object:

1. Open the 4-1_Using Leaders.dwg file.

2. Using the In-Canvas View Controls, restore the **Custom Model Views | 1-Add Leader to MTEXT** named view.

3. Using the **Home** ribbon and the **Annotation** panel, select the **Leader** command.

   First, we need to place the leader by Content.

4. Using the Command Line, key in C for the **Content First** command option.

   Next, we must select the MTEXT object to convert to a LEADER object.

5. Using the Command Line, key in M for the **Select Mtext** command option.

> **Note**
>
> Once the LEADER command is active, you can use the *right-click-slowly* mouse function to access the SELECT MTEXT command option.

ROOM 101

Figure 4.1: Adding a leader to MTEXT object results

6. Select the **ROOM 101** MTEXT object, move the cursor to define the symbol end of the leader line, and *left-click* to complete the command.

In the next exercise, I will demonstrate how to add and remove leaders without selecting a command.

Quickly remove a leader

Did you know there is a quicker way to add and remove leaders from an existing MULTI-LEADER object without using the ADD or REMOVE LEADER commands?

In this exercise, we will use the *Ctrl* key and the *Delete* key to modify a LEADER object.

1. Continue using the 4-1_Using Leaders.dwg file.

2. Using the In-Canvas View Controls, restore the **Custom Model Views | 2-Quick Edit Leaders** named view.

3. Use the *Ctrl* key and select one of the leader lines for the **701** LEADER object. The LEADER object will be displayed with HOT (magenta) grips. ■

4. Using the keyboard, press the *Delete* key to remove this leader.

Quickly modify a leader

Did you know that you can override the leader symbol, such as the arrowhead, using the same *Ctrl* key that you use to select the individual arrowheads?

Let's look at how to modify the leader symbol:

1. Continue using the 4-1_Using Leaders.dwg file.

2. Use the *Ctrl* key and select the CENTERLINE BENT LEADER, then *right-click* to open the **Properties** dialog.

3. Using the **Properties** dialog, locate the **Arrowhead** setting and use the drop-down list to select the **DOT Blank** symbol.

4. Use the *Esc* key to end the current command.

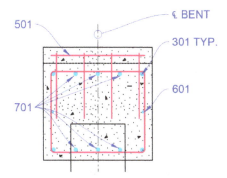

Figure 4.2: Modify leader symbol results

Adding multiple vertices to your leader

Did you know that you can use grips to add additional vertices to the LEADER object?

Let's look at how to add a leader vertex:

1. Continue using the 4-1_Using Leaders.dwg file.

2. Select the **601** LEADER object and *hover* over the **Symbol** grip to access the **Add Vertex** command.

3. Move the cursor to define the new symbol location, *left-click* to add the vertex, and move the symbol to the new location.

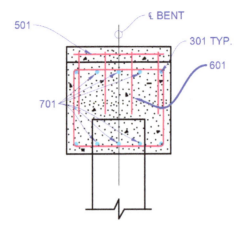

Figure 4.3: Add vertex results

> **Note**
>
> This is a common grip function when working with LEADER objects, so be sure to *hover* over all grips associated with the LEADER object to discover other commands that can be easily accessed to modify a LEADER object.

In the next exercise, we will look at how to modify the alignment of LEADER objects to improve the overall appearance of the drawing.

Making leaders parallel and evenly spaced

Many times, the LEADER objects are not visibly appealing and need to be re-aligned and distributed evenly to appear more organized. You can use the ALIGN LEADERS command to perform both functions:

1. Continue using the 4-1_Using Leaders.dwg file.
2. Using the In-Canvas View Controls, restore the **Custom Model Views | 3-Align Leaders** named view.
3. Using the **Annotate** ribbon and the **Leaders** panel, select the **Align Leaders** command.
4. Select all the LEADER objects and use the *Enter* key to complete the selection.
5. Using the Command Line, key in O for the **Options** command option.
6. Using the Command Line, key in D for the **Distribute** command option.

7. Select the BOTTOM LEADER object and drag the cursor up in a VERTICAL direction.

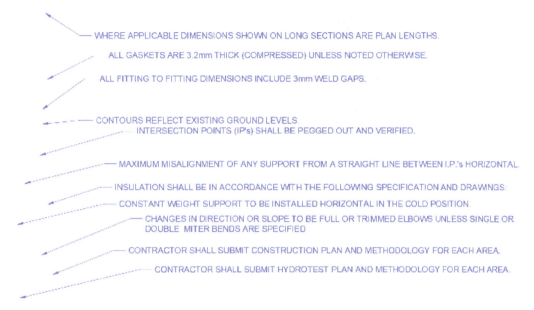

Figure 4.4: Align leaders' initial graphics

8. *Left-click* to define the line spacing as needed for these leaders.

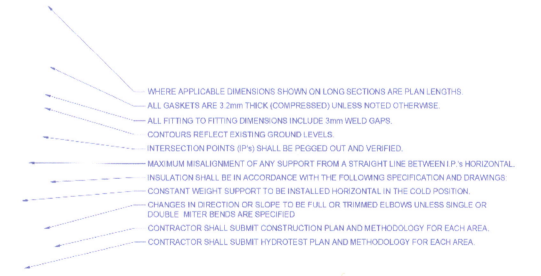

Figure 4.5: Align leaders results

You can also key in the value of the total line spacing, as I will demonstrate in the next set of instructions:

1. Continue using the `4-1_Using Leaders.dwg` file.

2. Using the In-Canvas View Controls, restore the **Custom Model Views | 4-Align by Dimension** named view.

3. Using the **Annotate** ribbon and the **Leaders** panel, select the **Align Leaders** command.

4. Select all the LEADER objects and use the *Enter* key to complete the selection.

5. Using the Command Line, key in O for the **Options** command option.

6. Using the Command Line, key in D for the **Distribute** command option.

7. Select the TOP LEADER object and drag the cursor down in a VERTICAL direction. Key in the value of 7 for the distance of the distribution, and use the Enter key to complete the command.

Figure 4.6: Align leaders with line spacing results

In the next section, we will look at how to modify and use hidden command options to use the linear dimension commands more efficiently.

Linear dimensions

Hopefully, you are using the All-in-One Dimension command for the majority of your linear dimensions. Let's review some of the command options.

All-in-One Dimension command

The new SMART Dimension command can significantly improve your use and placement of many dimensions, and it provides some handy command options to adjust dimensions on the fly. First, let's review the MOVE AWAY command option.

Move Away

The MOVE AWAY command option will automatically adjust your dimension lines to avoid conflicts:

1. Open the `4-2_All-in-One Dimensions.dwg` file.

2. Using the In-Canvas View Controls, restore the **Custom Model Views | 1-Move Away** named view.

3. Using the **Annotate** ribbon and the **Dimensions** panel, select the **Smart Dimension** command.

    When using the All-in-One dimension command, when the new dimension object falls on top of an existing dimension object, the MOVE AWAY command option will automatically adjust and move the existing overlapping dimension object away from the new dimension object.

    It helps if you are using **Dynamic Input** for this command.

4. Using the Status Bar, turn on **Dynamic Input**. Remember: blue is ON, gray is OFF.

5. *Hover* over the POLYLINE 2.0 TOP edge until the 2.0 linear dimension appears.

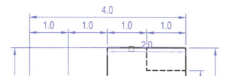

Figure 4.7: Preliminary dimension display

6. *Left-click* to select the TOP edge of the POLYLINE and drag the new dimension up until it is on top of the existing 4.0 dimensions. Then, *left-click* to accept this location.

    If you snap to the existing dimension line, a pop-up menu (with dynamic input enabled) offers these command options: **Move Away**, **Break Up**, **Replace**, or **None**. If you miss when selecting or hovering over the existing dimension, these options are not displayed, and the dimension is placed on top of the existing dimensions.

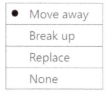

Figure 4.8: All-in-One Command options

7. Select **Move Away** from the pop-up menu and notice how the dimensions are adjusted to allow the new dimension to fit as needed.

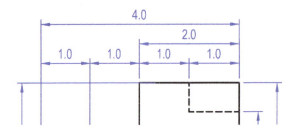

Figure 4.9: Move Away results

Break Up

The BREAK UP command option will split the existing dimension to create your new dimension:

1. Continue using the `4-2_All-in-One Dimensions.dwg` file.

2. Using the In-Canvas View Controls, restore the **Custom Model Views | 2-Break Up** named view.

3. Using the **Annotate** ribbon and the **Dimensions** panel, select the **Smart Dimension** command.

4. *Hover* over the POLYLINE TOP 2.0 line until the 2.0 linear dimension appears.

5. *Left-click* to select the TOP edge of the POLYLINE and drag the new dimension up until it is on top of the existing 4.0 dimensions. Then, *left-click* to accept this location.

6. Select **Break Up** from the pop-up menu and notice how the dimension is split and the 2.0 dimension is added.

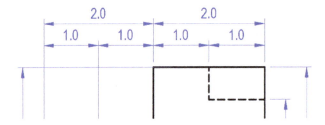

Figure 4.10: Break Up results

Replace

The REPLACE command option will replace the existing dimension with the new dimension:

1. Continue using the 4-2_All-in-One Dimensions.dwg file.

2. Using the In-Canvas View Controls, restore the **Custom Model Views | 3-Replace** named view.

3. Using the **Annotate** ribbon and the **Dimensions** panel, select the **Smart Dimension** command.

4. *Hover* over the POLYLINE TOP 2.0 line until the 2.0 linear dimension appears.

5. *Left-click* to select the TOP edge of the POLYLINE and drag the new dimension up until it is on top of the existing 1.0 dimensions, and *left-click* to accept this location.

6. Select **Replace** from the pop-up menu and notice how the 1.0 dimensions are replaced with the 2.0 dimension.

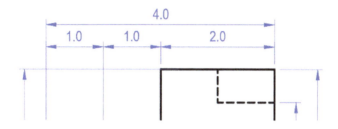

Figure 4.11: Replace results

Blank Dimensions

Have you ever wanted to create "blank" dimension graphics without having to edit out the text values? This can be accomplished by creating a dimension style that automatically suppresses the default text value:

1. Open the 4-3_LINEAR Dimensions.dwg file.

2. Using the In-Canvas View Controls, restore the **Custom Model Views | 1-BLANK Dimensions** named view.

3. Using the **Annotate** ribbon and the **Dimensions** panel, select the **Dimensions** dialog launcher icon ⌄ to open the **Dimension Styles** dialog.

4. Using the **Dimension Styles** dialog, select the **Standard** dimension style and click the **New** button.

5. Using the **Create New Dimension Style** dialog, change **New Style Name** to **Standard-BLANK** and click the **Continue** button.

6. Using the **New Dimension Style** dialog, selectsmart the **Primary Units** tab and change the **Prefix** value to \H, as shown here.

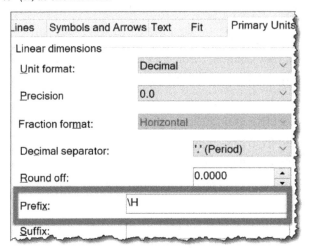

Figure 4.12: Blank dimension style

7. Click **OK**, then **Close** to close both dialogs.

8. Using the **Annotate** ribbon and the **Dimensions** panel, select the 4.0 dimension and *right-click* to access the **Dimension Style | Standard-BLANK** style.

Using this new style allows you to create the dimensions graphics without any text.

In the next exercise, we will learn how to use the TRIM and EXTEND commands on dimension objects.

Trimming and extending dimensions

Many users are unaware they can trim and extend dimension objects similar to other object types. Once you know you can use these commands on dimension objects, you will use them as often as I do.

Trimming dimensions

In this exercise, we will learn to TRIM the DIMENSION objects similar to trimming linear objects.

1. Continue using the 4-3_LINEAR Dimensions.dwg file.

2. Using the In-Canvas View Controls, restore the **Custom Model Views | 1-TRIM-EXTEND Dimensions** named view.

    The TRIM and EXTEND commands have changed after AutoCAD 2024. They now automatically trim or extend objects to any touching objects. You no longer need to select the boundary edges. While this is handy, it is very different from how these commands used to work.

To restore the previous functionality for both the TRIM and EXTEND commands, you can use the new MODE command option to restore the editing functionality to the STANDARD mode.

3. Using the **Home** ribbon and the **Modify** panel, select the **Trim** command.

4. Using the Command Line, key in O for the **Mode** command option and key in S to revert the command back to the previous **Standard** mode.

5. Use a crossing window to select all objects in the current view as the boundary edges and use the *Enter* key to complete the selection.

6. Select the **2.0 dimension** on the RIGHT half of the dimension line to trim the dimension back to the vertical dashed line.

7. Use the *Enter* key to complete the command.

Note

If you frequently use this Standard mode to access the previous TRIM command functionality, I recommend setting up a custom button for each TRIM mode: Standard and Quick.

Use the following key-ins for each of the following custom buttons:

- Trim-Standard: `^C^C_trim;O;S;\d\d;`

- Trim-Quick: `^C^C_trim;O;Q;\d\d;`

Extend dimensions

Next, we will use the TRIM command to EXTEND a dimension object:

1. Continue using the 4-3_LINEAR Dimensions.dwg file.

2. Using the In-Canvas View Controls, restore the **Custom Model Views | 1-TRIM-EXTEND Dimensions** named view.

3. Using the **Home** ribbon and the **Modify** panel, select the **Extend** command.

4. Use a crossing window to select all objects in the current view as the boundary edges and use the *Enter* key to complete the selection.

5. Select the **2.4 dimension** on the TOP half of the dimension line to extend the dimension to the TOP edge.

6. Use the *Enter* key to complete the command.

Note

You can use the *Shift* key to toggle the TRIM and EXTEND functionality in either the TRIM or EXTEND commands. In AutoCAD versions prior to 2024, you didn't have the MODE command option, so you could trim and extend dimensions much more easily. In AutoCAD 2024, trimming dimensions is not possible using the QUICK mode.

In the next exercise, we will learn how to add additional lines of text to a dimension without affecting the dimension line.

Multi-text dimensions

Often, we need to add text to our dimensions without losing the true dimension value, such as with TYP or other additional text. However, when you edit the text in a dimension and use the *Enter* key to add a second line of text, a portion of the dimension line is removed. That's not exactly what we want, right?

1. Continue using the 4-3_LINEAR Dimensions.dwg file.

2. Using the In-Canvas View Controls, restore the **Custom Model Views | 3-Multi-Text Dimensions** named view.

Figure 4.13: Initial multi-text dimension graphics

3. *Double left-click* on the text for the **4.0 dimension** to open the **TEXT Editor** dialog.

4. Click behind the dimension true value text currently displayed and key in \XTYP.

5. Use the *Tab* key to complete the modification in the **Properties** dialog.

6. *Left-click* in the view window to complete the text edit and close the **Text Editor** dialog.

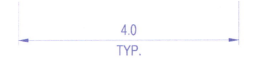

Figure 4.14: Multi-text dimension results

We would expect this result when using the *Enter* key to add additional text, right?

> **Note**
> You can also use the Properties dialog to add any additional text. <> designates the TRUE dimension value to remain.

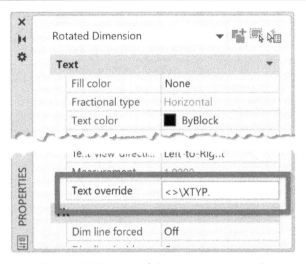

Figure 4.15: Properties | dimension text override

In the next exercise, we will learn how to fix those non-true dimension values in our drawing files, either singly or globally.

Resetting all dimension values to true values

Often, we run into a drawing file where someone has edited the dimension text to a value that is not the true value of that dimension. You don't know which dimension values have been edited , since they display the same as un-edited dimensions in the view window. There are two methods you can use to restore dimensions to their true values.

Method 1 – Express Tool | Reset Dimension

The first method for fixing a dimension value is to use the Express Tool | Reset Dimension:

1. Continue using the `4-3_LINEAR Dimensions.dwg` file.
2. Using the In-Canvas View Controls, restore the **Custom Model Views | 4-FIX Dimension Values** named view.
3. Select one of the dimensions and *right-click* to access the **Select Similar** command.
4. Using the Express Tools ribbon, locate the Dimension panel and select the Reset Text command. All dimension text values in the selection set are reset to their true dimension value.

The problem with using this method is that any additional text that has been added to a dimension is lost, such as the 4.23 TYP dimension in this file. This is not a desirable result, and we need to control this. A better way is to use an older method using the QSELECT command, which will only select dimensions that have been overridden. This way, you can isolate the overridden dimension objects and review them prior to running this command.

QSELECT	Command Locations
Ribbon	Home | Utilities | Quick Select
Command Line	QSELECT (QSE)
Right-click menu	Quick Select
Properties dialog	

Use *Ctrl + Z* to undo the previous modifications.

Method 2 – Properties dialog

The second method for fixing a dimension value is to use the **Properties** dialog to reset the dimension value:

1. Continue using the 4-3_LINEAR Dimensions.dwg file.
2.   Using the **Home** ribbon, locate the **Utilities** panel and select the **Quick Select** command.
3. Using the **Quick Select** dialog, change the **Object Type** to **Rotated Dimension**, or any/all dimension object types.
4. Change the **Operator** drop-down list to * **Wildcard Match**.
5. Using the **Value** field, key in ?* (this designates at least one character with a wildcard).
6. Verify that the setting for **Include in new Selection Set** is enabled and click **OK** to select the dimensions with overrides.

    Now that all the dimensions with overrides are selected, we can isolate them and be specific about which dimensions we want to restore to the true values.

7. Using the Status Bar, select the **Isolate Objects** icon and select the **Isolate Objects** command.

    If additional text is added to a dimension, we need to edit it using the text editor or the **Properties** dialog.

8. Select the **4.23 TYP** dimension and, using the **Properties** dialog, locate the **Text** panel and the **Text Override** field.
9. Modify the value to <>\XTYP. and use *Tab* or *Enter* on the keyboard to change the value.

In the next exercise, we will learn how to use an old command to get aligned dimensions that you just can't get with the DIMALIGNED command.

Forgotten DIMROTATED command

Have you ever had problems dimensioning objects that are staggered, or not aligned? Using the DIMALIGNED command doesn't place the dimension exactly the way you want, as shown here. So, what do you do?

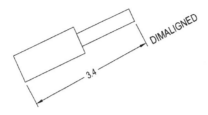

Figure 4.16: Initial DIMALIGNED dimension

Use the old DIMROTATED command to get exactly the dimension object you want.

DIMROTATED	Command Locations
Command Line	DIMROTATED (DIMRO)

In this exercise, we will learn to use the DIMROTATED command to place a staggered dimension.

1. Continue using the 4-3_LINEAR Dimension.dwg file.

2. Using the In-Canvas View Controls, restore the **Custom Model Views | 5-Using DIMROTATED** named view.

3. Using the Command Line, key in the DIMROTATED command.

4. Using the ENDPOINT osnap, snap to **P1** and **P2** to define the angle of the required dimension object.

5. Next, snap to the first extension line location, and then the second extension line location, using **P1** and **P3**.

6. Drag the dimension line to the preferred location and *left-click* to accept.

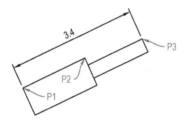

Figure 4.17: DIMROTATED results

In this section, we learned how to place and modify various linear dimension objects and their individual command options. After this section, we will be able to improve our productivity using linear dimensions.

In the next section, we will learn how to use the DIMSPACE command to clean up badly spaced linear dimensions.

DIMSPACE

We have all opened a drawing and seen badly spaced dimensions, right? Using the DIMSPACE command, you can easily fix the dimension spacing:

1. Continue using the 4-3_LINEAR Dimension.dwg file.
2. Using the In-Canvas View Controls, restore the **Custom Model Views | 6-Using DIMSPACE** named view.
3. Using the **Annotate** ribbon and the **Dimensions** panel, select the **Adjust Space** command.
4. Select the **4.0 dimension** as the BASE dimension.
5. Select all other dimensions as the dimensions to re-space and use the *Enter* key to complete the selection.
6. Use the *Enter* key to accept the **Auto** command option, or you can key in a distance to space the dimensions.

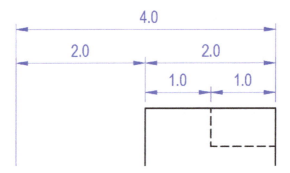

Figure 4.18: Initial DIMSPACE graphics

The DIMSPACE results appear as shown here.

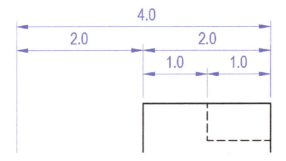

Figure 4.19: DIMSPACE results

In the next section, we will learn how to create angular dimensions that are larger than 180 degrees.

Angular dimensions

Do you need to create an angular dimension with angles larger than 180 degrees? Of course, we all do! If you use the Angular Dimension command and select the two lines, the only angles you get, by default, are angles less than 180 degrees. However, you can get the larger angles by using the command differently.

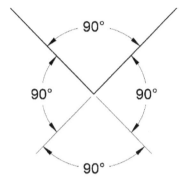

Figure 4.20: Angle larger than 180 degrees

Let's learn how to create angle dimensions greater than 180 degrees:

1. Open the 4-4_ANGULAR Dimensions.dwg file.
2. Using the In-Canvas View Controls, restore the **Custom Model Views | 1-Angle 180** named view.
3. Using the **Home** ribbon and the **Annotation** panel, select the expanded **Linear Dimension** command, and select the **Angular Dimension** command.
4. Instead of using the default command option to select the two lines, use the *Enter* key to use the command option to select the angle vertex.
5. Next, select the two angle vertices ENDPOINTS at **P1** and **P2**, and drag the angle dimension in the required direction and location.

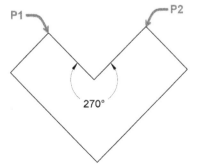

Figure 4.21: Angle larger than 180 degrees results

In the next section, we will learn how to use the new associated CENTERLINE and CENTERMARK commands.

Using smart centerlines

Have you taken advantage of the new CENTERLINE and CENTER MARK commands? Every once in a while, AutoCAD introduces a new feature that I didn't see coming, and this is one of them. I didn't even know I needed this functionality until it was here!

First, let's look at the CENTERLINE command.

Associated centerlines

When using the new CENTERLINE command, the centerline objects are associated with other objects, so when the objects are modified, the centerlines will maintain their association and automatically remain centered as needed:

1. Open the 4-5_Using CENTERLINES.dwg file.
2. Using the In-Canvas View Controls, restore the **Custom Model Views | 1-Using Centerlines** named view.
3. Using the **Annotate** ribbon and the **Centerlines** panel, select the **Centerline** command.
4. Select the TOP and BOTTOM lines of the POLYLINE shape to place the first CENTERLINE object.
5. Select the LEFT and RIGHT lines of the POLYLINE shape to place the second CENTERLINE object.

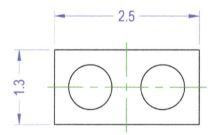

Figure 4.22: CENTERLINES results 1

Next, we will modify the POLYLINE shape to test the centerline association:

1. Use the *Esc* key to stop any current command.
2. Select the POLYLINE shape. Using grips, select the TOP MIDPOINT grip.

3. Drag the TOP edge of the POLYLINE up and key in 1 to extend the TOP edge up by one unit.

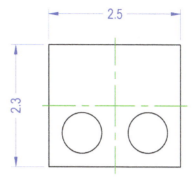

Figure 4.23: CENTERLINE results 2

Associated Center Marks

When using the new CENTER MARK command, the CENTER MARK objects are associated with the selected objects and automatically remain centered when the objects are modified:

1. Continue using the 4-5_Using CENTERLINES.dwg file.
2. Using the **Annotate** ribbon and the **Centerlines** panel, select the **Center Mark** command.
3. Select both CIRCLE objects to place their associated Center Marks.

Next, we will move the circles to verify that the center marks are associated:

1. Use the *Esc* key to stop any current command.
2. Select both CIRCLE objects, and using grips, select both **Center** grips.
3. Use the *Shift* key and select the **Center** grips on both CIRCLES.

> **Note**
>
> When using grips to edit an object, using the *Shift* key allows you to select more than one grips at a time.

4. Drag the grips UP and key in the value of 1. The Center Mark objects automatically follow the circle objects using association.

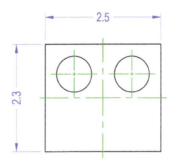

Figure 4.24: Center Marks results

The following system variables control how the CENTERLINE and CENTER MARK objects appear in the drawing file.

CENTERLAYER
This defines the default layer for new center mark and centerline object types.
Type: String
Saved In: Drawing
Initial Value: Use Current Layer (".")
The CENTERLAYER system variable applies to center marks and centerlines created using the CENTERMARK and CENTERLINE commands only. Any value other than "." (current layer) will override the Current Layer (CLAYER) system variable.

CENTEREXE
This defines the length of the centerline extensions for center marks and centerline object types.
Type: Real
Saved in: Drawing
Initial value: 0.1200 (imperial) or 3.5000 (metric)
The CENTEREXE system variable applies to the length of centerline extensions created using the CENTERMARK and CENTERLINE commands only. This value can only be positive real numbers.

CENTERLTYPE
This defines the linetype used for center marks and centerline object types.
Type: String
Saved in: Drawing
Initial value: CENTER2
The CENTERLTYPE system variable applies to the linetype for center marks and centerlines created with the CENTERMARK and CENTERLINE commands only. The following values are accepted: "." (use current), ByLayer, or any valid linetype name.

"."	Use Current Layer (CLAYER)
ByLayer	
Any valid linetype name	

CENTERLTSCALE
This defines the linetype scale used by center marks and centerline object types.
Type: Real
Saved in: Drawing
Initial value: 1.0000
The CENTERLTSCALE system variable applies to center marks and centerlines created with CENTERMARK and CENTERLINE commands only. This system variable can use any real number except zero.

CENTERMARKEXE
This defines whether the centerlines created using the centerline command extend automatically from center marks.
Type: Switch (on/off)
Saved in: Drawing
Initial value: 1
The CENTERMARKEXE system variable applies to center marks and centerlines created with CENTERMARK and CENTERLINE commands only.

CENTERCROSSSIZE
This defines the size of the associative center mark object.
Type: String
Saved in: Drawing
Initial value: 0.1x
The CENTERCROSSSIZE system variable applies to center marks created using the CENTERMARK commands only. The CENTER MARK default size is 0.1x with a gap of 0.05x, which is controlled using the CENTERCROSSGAP system variable.

CENTERCROSSGAP
This defines the size of the gap between the center mark object and its centerlines.
Type: String
Saved in: Drawing
Initial value: 0.05x
The CENTERCROSSGAP size is ignored when the radius of the circle is less than the sum of half the center mark size and the gap distance. This problem can be avoided by specifying relative values for both CENTERCROSSSIZE and CENTERCROSSGAP system variables.

CENTERDISASSOCIATE	Command Locations
Command Line	CE
This command removes the association between center mark and centerline objects	

CENTERREASSOCIATE	Command Locations
Command Line	CENTERR
This command adds an association between center mark and centerline objects	

Bonus commands

In this section, we will learn some of the lesser-known dimension commands, such as DIMLAYER and DIMBREAK, to control our dimension objects. We will also learn how to easily control how they work.

DIMLAYER

Use the DIMLAYER system variable to control automatically which layer is used for all dimensions.

DIMLAYER	
This defines the default layer for new dimensions.	
Type:	String
Saved in:	Drawing
Initial value:	"." (use current)
Layer name values other than "." (use current) will override the current layer (CLAYER).	

DIMBREAK

The DIMBREAK command allows you to "break" the dimension or extension lines where they cross other objects. With it, you no longer need to "explode" any dimensions to clean up those crossing objects. Don't worry; it doesn't really break the dimension, it just fakes it out!

Dimensions

In this exercise, we will learn to use the DIMBREAK command to break object lines to improve readability of the drawing.

1. Open the 4-6_Bonus DIMENSION Commands.dwg file.
2. Using the In-Canvas View Controls, restore the **Custom Model Views | 1-Using DIMBREAK** named view.
3. Using the **Annotate** ribbon, locate the **Dimensions** panel and select the **Break** command.
4. Select the horizontal extension line at **P1** as the object to break.
5. Select the vertical extension line at **P1** as the object to use for the break.

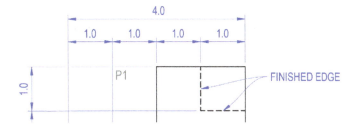

Figure 4.25: DIMBREAK linear dimensions 1

You can also perform this break on multiple objects:

1. Use the *Spacebar* to recall the previous **Dimbreak** command.
2. Using the Command Line, key in M to use the **Multiple** command option.
3. Use a crossing window to select all the horizontal dimensions.
4. Again, use a crossing window to select all the vertical dimensions.
5. Use the *Enter* key to accept the default **Auto** command option.

All dimension extension lines that overlap are now cleaned up for cleaner-looking drawings.

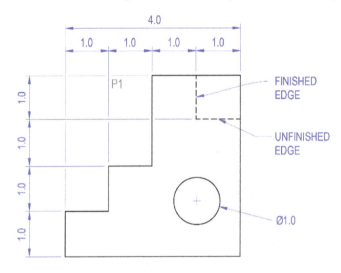

Figure 4.26: DIMBREAK linear dimensions 2

You can also clean up the radial dimension:

1. Use the *space bar* to recall the previous **Dimbreak** command.
2. Select the **Radius** dimension.
3. Select the vertical object line that overlaps the **Radius** dimension.

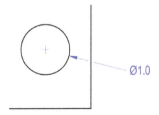

Figure 4.27: DIMBREAK radial dimension

Leaders

You can break LEADER objects, as well as other dimension objects, as long as you use the STRAIGHT leader type. For some unknown reason, the SPLINE leader type cannot be broken using the DIMBREAK command:

1. Continue using the 4-6_Bonus DIMENSION Commands.dwg file.
2. Using the **Annotate** ribbon and the **Dimensions** panel, select the **Break** command.
3. Select the UNFINISHED EDGE leader with the **Straight** leader type.
4. Select the vertical line as the object to use for the break.

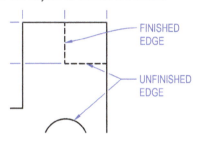

Figure 4.28: DIMBREAK leaders results

> **Note**
>
> DIMBREAK will not work on LEADER objects using the SPLINE leader type or the older QLEADER object type. You can use the LEADERTOMLEADER utility from the AutoCAD App Store to convert older LEADER objects to MLEADER objects.

> **Note**
>
> DIMBREAK will not work on the leader symbols or on the leader text. If the dimension or leader symbol is moved, the break will be applied. You can use a WIPEOUT object to mask a dimension or leader symbol.

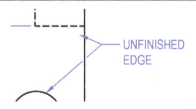

Figure 4.29: DIMBREAK at leader symbol

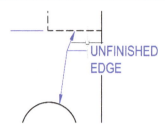

Figure 4.30: BACKGROUND MASK at leader text

> **Note**
>
> Use a BACKGROUND MASK to mask the LEADER TEXT in a LEADER object. How to use a BACKGROUND MASK is discussed in *Chapter 3*.

Blocks and references

The DIMBREAK command cannot be used on dimensions in BLOCKS or REFERENCES, but the objects in a BLOCK or REFERENCE can be used as the cutting edge for dimensions in the current file:

1. Continue using the 4-6_Bonus DIMENSION Commands.dwg file.

2. Using the In-Canvas View Controls, restore the **Custom Model Views | 2-XREF and DIMBREAK** named view.

3. Using the **Annotate** ribbon and the **Dimensions** panel, select the **Break** command. ⊥⊣

4. Select the vertical extension line at **P1** as the object to break.

5. Select the horizontal extension line (in the XREF) at **P1** as the object to use for the break.

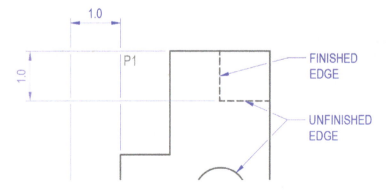

Figure 4.31: DIMBREAK XREF dimensions results

> **Note**
> You can use the REMOVE command option to remove a break when necessary.

The following objects can be used as cutting edges when using the DIMBREAK dimension command:

- Leader
- Line
- Blocks (limited to the previously mentioned rules)
- Xrefs (limited to the previously mentioned rules)
- Spline
- Ellipse
- Polyline
- Text
- Multiline text
- Circle
- Arc

> **Note**
> You can control the size of the dimension breaks using Annotation Scale for the current viewport.

In the next exercise, we will learn how to control the size and scale of our dimensions in multiple viewports.

Summary

In this chapter, we learned how to place various dimension types more quickly and how to control the appearance of our dimensions so they follow more standard drafting practices. We also learned how to define various system variables to control how the dimension commands work and how to automatically control the dimension CAD Standards using system variables.

In the next chapter, we will look at how to work with the TABLE object to gain smarter tables. We will also look at how to create multiple table types.

5

Making Tables Work for You

In this chapter, you will learn how to use the TABLE object to its full potential. TABLE objects are used to gather data into schedules using various data types from drawing objects. If you are still manually collecting data from your AutoCAD drawing files, manipulating it in another application, and then importing it back into AutoCAD, you need to read this chapter to learn how to automate these processes and discover all the possibilities when working with TABLE objects.

In this chapter, we'll cover the following topics:

- Working with tables and cells

- Creating smarter tables

- Using table links

- Using table and cell styles

- Automating the use of tables with Tool Palettes

- Bonus table commands

By the end of this chapter, you will be able to take advantage of smarter data manipulation using TABLE objects, including formulas, fields, and table linking.

Working with Tables And Cells

I assume you know the basics of selecting and creating a simple table and can navigate and add rows and columns. But are you familiar with some of the more specific table features such as locking cells, copying and incrementing values, using formulas or fields in a table, or auto-fill cells?

First, I want to review the available grips when working on a TABLE object.

Using Table Grips

First, let's look at the common grips available when a table is selected:

1. Use these grips ■ to modify the height of a single row.

2. Use these grips ■ to modify the width of a single cell.

3. Use this grip ◆ to increment the value of a cell automatically using the options, such as Fill Series and Copy Cells with or without formatting.

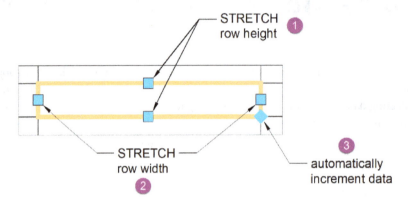

Figure 5.1: Using Table Grips

In this exercise, we will look at how to copy and increment cell data within a TABLE object.

Copying and Incrementing Cell Data

I assume you know the basics of selecting and creating a simple table and can navigate within a table to add rows and columns. But are you familiar with selecting multiple cells, rows, or columns using common AutoCAD selection tools, such as a crossing window for a formula in a table?

In this exercise, we will learn how to use the auto-increment feature to fill in multiple table cells:

1. Open the `5-1_Working with Tables.dwg` file.

2. Using the In-Canvas View Controls, restore the **Custom Model Views | 1-Auto Increment Data** named view.

3. Select the TABLE object by selecting the edge of the table and issue a *left-click* "inside" the **A3** cell.

4. Select the **Auto-Increment** grip ◆ and drag it to the bottom of the **A** column. This will copy the value of the **A3** cell to all cells in the **A** column.

	A	B	C
1	SCHEDULE		
2	NAME	NUMBER	QUANTITY
3	D101	100	1000
4	D101		
5	D101		
6	D101		
7	TOTALS		

Figure 5.2: Auto-Increment 1 Results

5. Select the TABLE object by issuing a *left-click* "inside" the **B3** cell.

6. Select the **Auto-Increment** grip ◆ and drag it to the bottom of the **B** column. This will copy and increment the value of the **B3** cell to all cells in the **B** column.

	A	B	C
1	SCHEDULE		
2	NAME	NUMBER	QUANTITY
3	D101	100	1000
4	D101	101	
5	D101	102	
6	D101	103	
7	TOTALS		

Figure 5.3: Auto-Increment 2 Results

> **Why are these grips different?**
>
> The INCREMENT AND COPY command only increments the value if the data format is defined as decimal number, angle, percentage, and currency. All other data formats perform the COPY command.

You can avoid the increment of number fields using the following steps:

1. Select the TABLE object by issuing a *left-click* "inside" the **C3** cell.

2. Select the **Auto-Increment** grip ◆ and *right-click* to access the following command options:

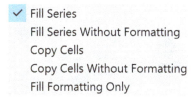

Figure 5.4: Auto-Increment Command Options

3. Select the **Copy Cells** option and drag the grip to the bottom of the **C** column. This will copy the value of the **C3** cell (no increment) to all cells in the **C** column.

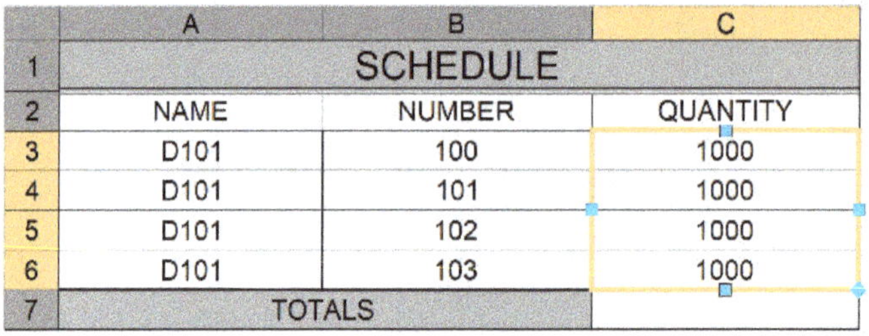

	A	B	C
1	SCHEDULE		
2	NAME	NUMBER	QUANTITY
3	D101	100	1000
4	D101	101	1000
5	D101	102	1000
6	D101	103	1000
7	TOTALS		

Figure 5.5: Auto-Increment Results 3

In the next exercise, we will learn how to lock specific data formats and values in our tables.

Locking a Table Cell

Use the LOCK CELL command options to prevent cell modifications. You can lock the cell properties for content, format, or both:

1. Continue using the `5-1_Working with Tables.dwg` file.

2. Select the TABLE object by issuing a *left-click* "inside" the **A3** cell.

> **Note**
>
> *Hover* over the TABLE object to preview the column, row, and cell numbers.

3. Using the **Table Cell** ribbon and the **Cell Format** panel, hold down the **Cell Locking** button and select the **Content and Format Locked** command.

> **Note**
>
> The **Table Cell** ribbon is a context sensitive ribbon and only appears when a TABLE object is selected.

A LOCK icon 🔒 and tooltip will appear on the cursor when you *hover* over a selected and "locked" cell in a table.

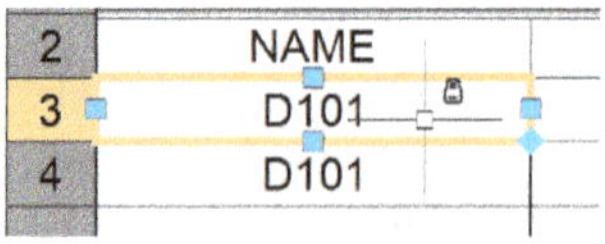

2	NAME
3	D101
4	D101

Figure 5.6: Locked Cell in a Table

Unlocking a Table Cell

Use the UNLOCK CELL command to remove a table cell locked state:

1. Select the TABLE object by selecting "inside" the **A3** cell.

2. Using the **Table Cell** ribbon and the **Cell Format** panel, hold down the **Cell Locking** button and select the **Unlocked** command.

> **Note**
>
> With a table cell selected, use the *right-click* to access the **Lock** and **Unlock** commands.

In the next exercise, we will look at how to use formulas to calculate data in a table.

Using a Total Formula

In this exercise, we will learn how to select multiple cells in a column to easily create a Total column using a formula:

1. Continue using the `5-1_Working with Tables.dwg` file.

2. Using the In-Canvas View Controls, restore the **Custom Model Views | 2-Total Formula** named view.

3. Select the TABLE object by issuing a *left-click* inside the **A8** cell.

4. Hold down the *Shift* key and *left-click* "inside" the **B8** cell. This will select both cells simultaneously.

5. Using the **Table Cell** ribbon and the **Merge** panel, select the **Merge Cells** *drop-down list* and select the **Merge All** command 🔲.

6. Select the newly merged cell and key in the value TOTALS in the newly merged cell, **A8:B8**.

    Next, we need to add SUM formulas to cells C8 and D8.

7. *Left-click* "inside" the **C8** cell.

8. Using the **Table Cell** ribbon and the **Insert** panel, select the **Formula** *drop-down list* and select the **Sum** command.

9.   *Left-click and drag* a crossing window over cells **C3** to **C7**.

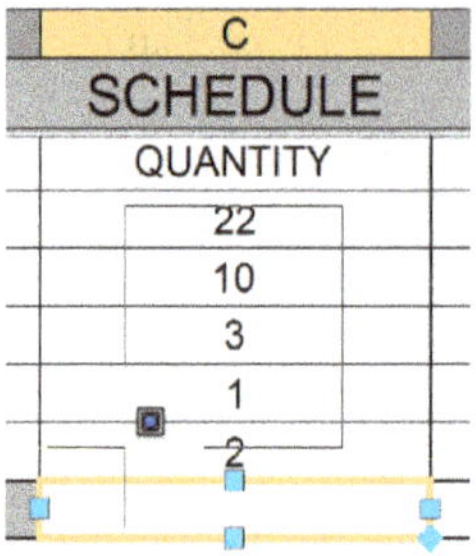

Figure 5.7: Crossing Selection Cells

10. The formula =**SUM(C3:C7)** is placed in the **C8** cell. *Left-click* anywhere in the view window to complete the command.

11. *Left-click* inside the FOURTH column, BOTTOM cell, or the **D8** cell.

12. Using the **Table Cell** ribbon and the **Insert** panel, select the **Formula** *drop-down list* and select the **Sum** command.

13. *Left-click and drag* a CROSSING window over cells **D3** to **D7**.

14. *Left-click* anywhere in the view window to complete the command. The formula =**SUM(D3:D7)** is placed in the **D8** cell.

SCHEDULE				
NAME	SIZE	QUANTITY	COST	TOTAL COST
D30	2'8" X 7'	22	$750.00	
D36	3' x 7'	10	$1250.00	
D48	4' x 7'	3	$1500.00	
D36-8	3' x 8'	1	$1750.00	
D48-8	4' x 8'	2	$2500.00	
TOTALS		38	$7750.00	

Figure 5.8: Total SUM Columns

Using a Total Equation

Next, we want to create the formulas to calculate the Quantity x Cost = Total Cost for each cell in the **E** column:

1.   Continue using the `5-1_Working with Tables.dwg` file.

2.   *Left-click* inside the LAST column, FIRST row cell, or the **E3** cell.

3.   Using the **Table Cell** ribbon and the **Insert** panel, select the **Formula** *drop-down list* and select the **Equation** command $f_{(x)}$.

4. Key in the formula =C3*D3 in the **E3** cell.

5. *Left-click* anywhere in the view window to complete the command.

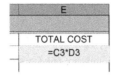

Figure 5.9: Total Cost Equation

6. Use *Ctrl + C* (copy to clipboard) to copy the formula in cell **E3**.

7. Use *Ctrl + P* (paste from clipboard) to paste the formula into cells **E4** to **E7**.

Note

Use the *Shift* key to select the FIRST and LAST cell to select all cells in between.

Note

If a formula error occurs, the table displays the hashtag (#) in the cell.

The errors display as hashtags when there is a formula error.

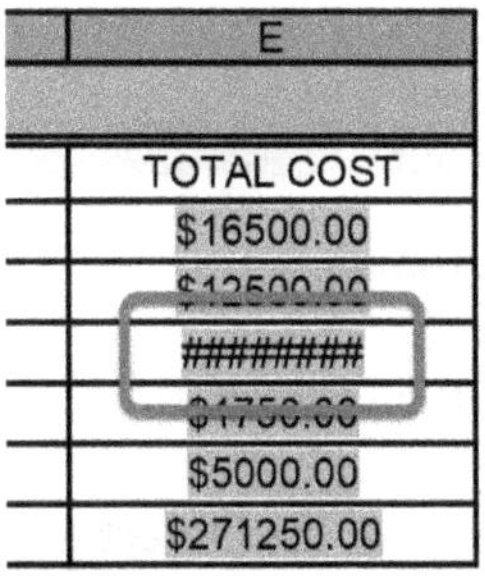

E
TOTAL COST
$16500.00
$12500.00
#######
$1750.00
$5000.00
$271250.00

Figure 5.10: Error Display Example

In this section, we learned how to automatically copy and increment values in a table cell, how to lock and unlock cells, and how to add formulas to a table in AutoCAD.

In the next section, we will look at how to create a Legend or Symbol table.

Creating Smarter Tables

You can insert FIELDS and BLOCKS into a table cell to create smarter tables for your project, such as Legend or Schedule tables.

Creating a Table Using Fields

In this exercise, we will populate a table cell with a field that retrieves its value from the length of the rebar:

1. Open the 5-1_Working with Tables.dwg file.
2. Using the In-Canvas View Controls, restore the **Custom Model Views | 3-Rebar Lengths** named view.
3. Select the TABLE object by issuing a *left-click* "inside" the **C5** cell.
4. Using the **Table Cell** ribbon and the **Insert** panel, select the **Field** command 📇.
5. Using the **Field** dialog, define the following settings:

   - Field category: **Objects**
   - Field names: **Object**

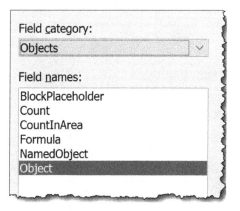

Figure 5.11: Field Settings 1

6. Select the **Click Object** button 🔳 for the **Object Type** and select the RED REBAR (501) object in the view window.
7. Using the **Field** dialog, define the remaining settings:

   - Property: **Length**
   - Format: **Architectural**

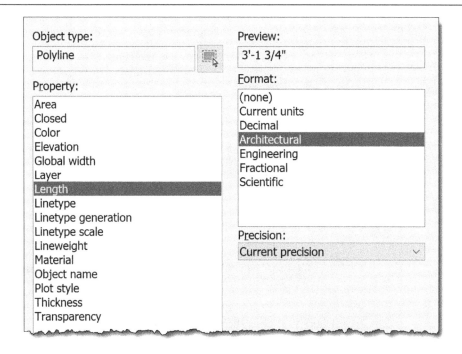

Figure 5.12: Field Settings 2

8. Click **OK** to save these changes and close the dialog.

9. Press the *Enter* key to complete the command.

Estimated Quantities			
BAR	**NO.**	**LENGTH**	**TOTAL LENGTH**
301	4	44'-8"	
401	8	4'-4"	
501	20	2'-0"	
601	2	4'-7"	
701	6	44'-8"	
	TOTAL		

Figure 5.13: Adding Field to Table Cell

The table cell is populated with the **2'-0"** length from the selected rebar in the drawing.

10. Select the RED 501 REBAR line and use grips to modify the length of the rebar object.

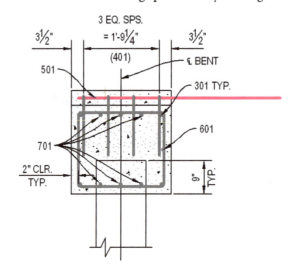

Figure 5.14: Modify Rebar 501 Length

11. Using the Command Line, key in the REGEN command to force the FIELD objects in the table to update.

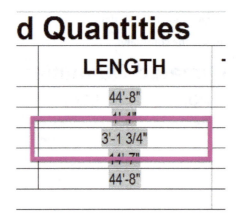

Figure 5.15: REBAR 501 Length Results

You can control the update methods for fields using the following system variable.

FIELDEVAL	
Controls how fields are updated. This setting is stored as a bitcode using the sum of the following values:	
Type:	Integer
Saved in:	Drawing
Initial Value: 31	
0	At Least. Adjusts line spacing based on the tallest characters in a line.
1	Exactly. Uses the specified line spacing, regardless of individual character sizes.
2	Not updated
4	Updated on open
8	Updated on save
16	Updated on plot
31	Updated on the use of ETRANSMIT

> **Note**
>
> You can update any individual FIELD object(s) using the **Update Field** command and selecting objects.

UPDATE FIELD	Command Locations
Ribbon	Insert | Data | Update Fields
Command Line	UPDATEFIELD (UPDA)

In the next section, we will learn to create links between drawing objects and an external Excel spreadsheet.

Creating a Legend Table

First, we need to add blocks to a TABLE object to create a Legend table. The size and appearance of the block can be set automatically or defined manually. Multiple blocks can be inserted using the Manage Cell Content dialog.

Inserting a Block

In this exercise, we will learn to insert a drawing BLOCK object into a cell in the TABLE object.

1. Open the 5-2_Create a Legend Table.dwg file.
2. Using the In-Canvas View Controls, restore the **Custom Model Views | 1-Legend Table** named view.

3. Select the TABLE object by issuing a *left-click* "inside" the **A2** cell.

4. Using the **Table Cell** ribbon and the **Insert** panel, select the **Block** command.

5. Using the **Insert a Block in a Table Cell** dialog, select the block's **Name** *drop-down-list* and select the **North_Arrow** block.

6. Turn OFF the setting for **AutoFit**.

7. Modify **Overall cell alignment** to **Middle Center**.

8. Click **OK** to save the changes and close the dialog.

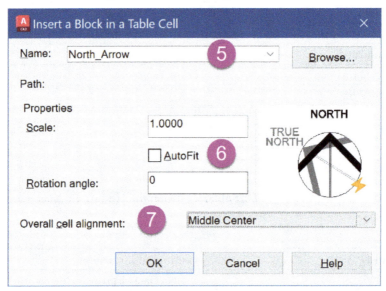

Figure 5.16: Inserting a Block into a Cell

9. Repeat *steps 4–8* to add the following blocks to the Legend table:

   • North Arrow

   • Graphic Scale-Imperial

   • Revision-TRIANGLE

   • Elevation

   • Section

   • Keynote

   • Title

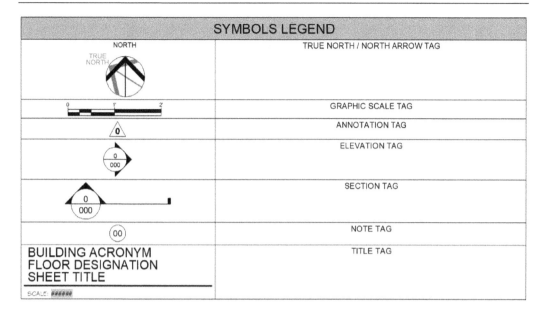

Figure 5.17: Legend Table Results

Inserting Multiple Blocks

You can add more than one block to a single table cell using the following steps:

1. Continue using the `5-2_Create a Legend Table.dwg` file.
2. Select the TABLE object by issuing a *left-click* "inside" the **A4** cell, Annotation Tag.
3. Using the **Table Cell** ribbon and the **Insert** panel, select the **Block** command ⊞.
4. Using the **Insert a Block in a Table Cell** dialog, select the block's **Name** *drop-down list* and select the **Tag-DIAMOND** block.
5. Turn OFF the setting for **AutoFit**.
6. Modify **Overall cell alignment** to **Middle Center**.
7. Click **OK** to save the changes and close the dialog.
8. Repeat *steps 2–6* to add the **Tag-HEX** block to the **A4** cell.
9. Using the **Table Cell** ribbon and the **Insert** panel, select the **Manage Cell Contents** command.
10. Use the **Manage Cell Contents** dialog to rearrange or remove the blocks in a single table cell.

> **Note**
>
> The **Manage Cell Contents** command is only available when two or more items are inserted into a single cell.

11. Using the **Table Cell** ribbon and the **Cell Styles** panel, select the **Alignment | Middle Center** command to re-align the blocks in the table cell.

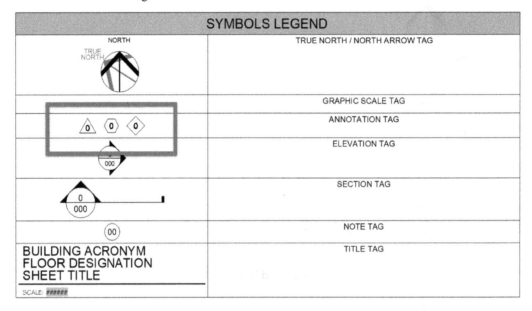

Figure 5.18: Adding Multiple Blocks in a Table Cell

In the next exercise, we will use text fields to populate table cells using automatic values extracted from objects within the drawing.

Using Table Links

A TABLE object can be linked to data outside of AutoCAD if it is stored in a Microsoft Excel (XLS, XLSX, or CSV) file. The table can be linked to the entire spreadsheet, an individual row, a column, a cell, or a cell range in Excel.

This linked data can be linked to the AutoCAD table using the following three methods:

- As formulas with data formats attached
- As calculations from formulas calculated in Excel (with data formats not attached)
- As calculations from formulas calculated in Excel (with data formats attached)

A table that contains linked data will display "link indicators" around the linked cells. Information about the linked data can be viewed by "hovering" over the data link "indicators."

Linking a Table to Excel

In this section, we will create a new table that reads block attributes from the active and referenced drawings and links an attribute value to an external Excel spreadsheet:

1. Open the 5-3_Using Table Links.dwg file.

2. Using the In-Canvas View Controls, restore the **Custom Model Views | 1-Link Table to Excel** named view.

3. Select the desk in the conference room and *right-click* to access the **Edit Xref In Place** command.

4. Using the **Reference Edit** dialog, verify that the **Automatically Select All Nested Objects** setting is activated and click **OK** to close the dialog, and then open the reference file for editing.

5. *Double left-click* on the conference room desk to review the associated attributes.

6. After reviewing the attached attributes, click the **Cancel** button to close the dialog.

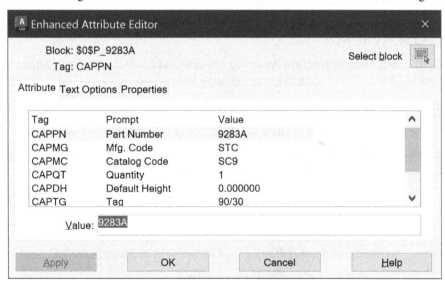

Figure 5.19: Desk Block Attributes

The spreadsheet data includes the following data:

	Name	CAPPN
1		
2	P_2006	2006
3	P_2036-WLAM	2036-WLAM
4	P_287	287
5	P_421482N	421482N
6	P_462LEAP9L	462LEAP9L
7	P_50622625	50622625
8	P_730581	730581

Figure 5.20: Excel Spreadsheet Data

7. Using the active **Insert** ribbon and the **Edit Reference** panel, select the **Discard Changes** command and click **OK** to complete the **Edit Xref In-Place** command.

Next, we want to use the Part Number (CAPPN) block attribute in the AutoCAD block to link to the identical column in the Excel spreadsheet. This link will allow us to include information from the spreadsheet in our AutoCAD TABLE object.

Linking to the PART NUMBER (CAPPN)

First, we want to place this table in Paperspace, so we need to toggle to the Paperspace environment:

1. Using the Model Tabs above the Status Bar, select the **My Layout** Tab.
2. Using the **Annotate** ribbon and the **Tables** panel, select the **Table** command ⊞.
3. Using the **Insert Table** dialog, select the **From Object Data In The Drawing (Data Extraction)** setting to create a table and click **OK** to continue.
4. Using the **Data Extraction-Begin (Page 1 of 8)** dialog, select the **Create a new data extraction** option and click the **Next >** button.
5. Using the **Save Data Extraction As** dialog, navigate and create the new data extraction file (`FURNITURE.DXE` or `.DXEX`) in the following folder:

    `...ACAD_TipsTechniques\Exercise_Files\Chapter 05\FURNITURE.DXEX`

6. Click the **Save** button to save the new `FURNITURE.DXEX` file.
7. Using the **Data Extraction-Define Data Source (Page 2 of 8)** dialog, click the **Settings** button and turn ON the **Include Xrefs In Block Counts** option.
8. Click **OK** to close the **Data Extraction – Additional Settings** dialog and continue.

> **Note**
> At this stage, you can add additional drawings and folders to be included in the TABLE object if needed.

9. Click the **Next >** button to read all objects from the selected files.
10. Using the **Data Extraction-Select Objects (Page 3 of 8)** dialog, turn OFF the **Display all object types** setting.
11. Turn ON the **Display blocks with attributes only** setting.
12. Clear the "checkmark" for all blocks not named **P_***. You can simplify this by selecting the **Object** column to re-sort the blocks based on their name.

> **Note**
>
> *Right-click* anywhere in this list to access the shortcuts: Check All, Uncheck All, Invert Selection, and Edit Display Name to easily select all blocks in the list.

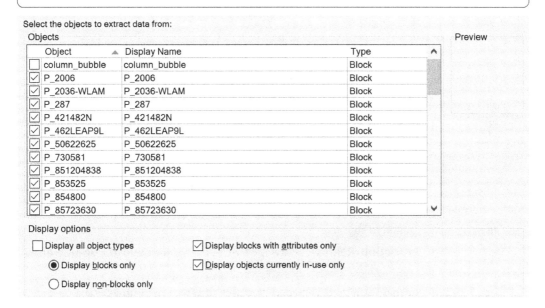

Figure 5.21: Selecting Blocks to Extract

13. Click the **Next >** button to continue and turn OFF all Properties except **CAPMC** and **CAPPN**.

14. Using the **Data Extraction-Select Properties (Page 4 of 8)** dialog, locate the **Category Filter** section, and uncheck all categories except **Attribute**.

> **Note**
>
> *Right-click* on the **Category filter** list to quickly Check All or Uncheck All categories.

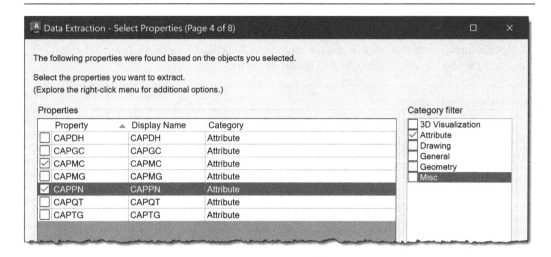

Figure 5.22: Defining Data Extraction Properties

15. Click the **Next >** button to continue.

16. Using the **Data Extraction-Refine Data (Page 5 of 8)** dialog, turn OFF the **Show Name Column** setting (it is duplicated in the spreadsheet data).

17. Click the **Link External Data** button and click the **Launch Data Link Manager** button.

18. Using the **Data Link Manager** dialog, select **Create a new Excel Data Link**.

19. Using the **Enter Data Link Name** dialog, key in the name Furniture and click **OK** to create the data link.

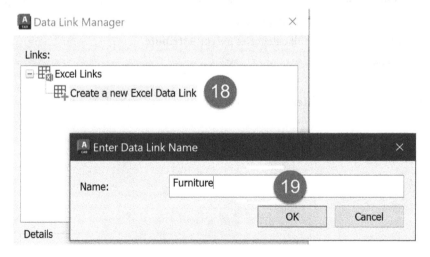

Figure 5.23: Creating a Data Link

20. Using the **New Excel Data Link: Furniture** dialog, select the **Browse For A File | Browse** button ▪▪▪ and navigate to the following spreadsheet location:

    `...\ACAD_TipsTechniques\Exercise_Files\Chapter 05\FURNITURE.XLS`

21. Click **OK** to close the dialog, and click **OK** again to close the **Data Link Manager** dialog.

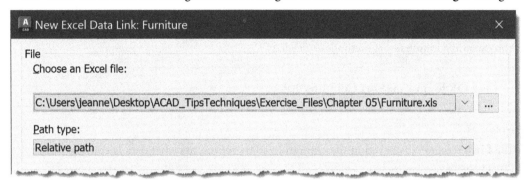

Figure 5.24: Spreadsheet File Location

Next, we need to identify the CAPPN data value as the "link key" between the drawing file and the Excel spreadsheet.

1. Using the **Link External Data** dialog, select Furniture for the source of the external data.

2. Select **CAPPN** for **Drawing data column,** and **CAPPN** for **External data column**.

3. Click the **Check Match** button to verify that each file has the associated data name.

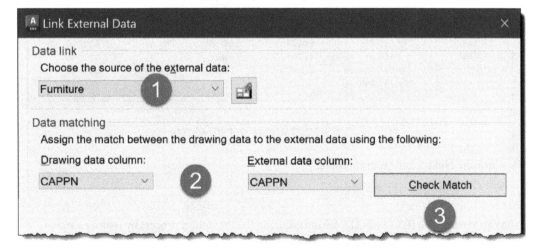

Figure 5.25: Defining a Data Link Value

4. If successful, you will see a **Valid Key** dialog appear stating **The Key Pairing Was Successful**.

5. Click **OK** to close the **Valid Key** dialog and click **OK** to close the **Link External Data** dialog.

6. Using the **Data Extraction-Refine Data (Page 5 of 8)** dialog, click the **Sort Columns Options** button and use the **Select Column** *drop-down list* to select **CAPPN** to sort in **Ascending** order.

> **Note**
>
> You can add additional sort methods as needed.

7. Click **OK** to close the **Sort Columns** dialog.

> **Note**
>
> The **Link** icon displayed in the column headers represents what data is coming from the spreadsheet.

8. Using the **Data Extraction-Refine Data (Page 5 of 8)** dialog, *drag and drop* the column headers to re-arrange the columns as shown in the following figure. Shrink the columns as much as you can to simplify this *drag-and-drop* arrangement. Yes, it is tedious, to say the least. I sure wish we could get an

Name - (External)	CAPPN	CAPMC	CAPTG	Count	COST
P_9243A	9243A	SC9	75/25	1	$112.00
P_981830T	981830T	SC9	TB/30	1	$118.00
P_GCTR4848	GCTR4848	SCW	48	1	$113.00
P_99995	99995	SC9	30/45/45/30	1	$128.00
P_98013P3	98013P3	SCT	30/42	1	$117.00
P_2036-WLAM	2036-WLAM	BEC	36	1	$101.00
P_GCV72	GCV72	SCW	SM/72	1	$114.00
P_CLEJ3660	CLEJ3660	SCW	60/36	1	$107.00
P_854800	854800	SCA	48	1	$109.00
P_GCD7236	GCD7236	SCW	72/36	2	$110.00
P_GCN7224PKP	GCN7224PKP	SCW	72/24	2	$111.00

Figure 5.26: Arranging Data Extraction Columns

9. Click the **Next >** button to continue.

> **Note**
>
> You can rename the (EXTERNAL) columns during the extraction process by issuing a *right-click* on any column header to access the **Rename** column command.

10. Using the **Data Extraction-Choose Output (Page 6 of 8)** dialog, turn ON the **Insert Data Extraction Table Into Drawing** setting and click the **Next >** button to continue.

11. Using the **Data Extraction-Table Style (Page 7 of 8)** dialog, key in the value FURNITURE SCHEDULE for the title of the new table and click the **Next >** button to continue.

12. Using the **Data Extraction-Finish (Page 8 of 8)** dialog, click the **Finish** button to close the dialog.

13. Use **ENDPOINT** to snap to the LOWER-LEFT corner of the viewport to place the table on the sheet layout.

Name - (External)	CAPPN	CAPMC	CAPTG	Count	COST
P_853525	853525	SCA	35/20	1	$108.00
P_GCTN7236	GCTN7236	SCW	72/36	1	$112.00
P_GCB4224	GCB4224	SCW	42/24	1	$109.00
P_85723630	85723630	SCA	72/36	1	$110.00
P_9243A	9243A	SC9	75/25	1	$112.00
P_981830T	981830T	SC9	TB/30	1	$118.00
P_GCTR4848	GCTR4848	SCW	48	1	$113.00
P_99995	99995	SC9	30/45/45/30	1	$128.00
P_98013P3	98013P3	SCT	30/42	1	$117.00

FURNITURE SCHEDULE

Figure 5.27: Placing the Data Extract Table on a Layout

In this exercise, we learned how to link and generate a table using block attributes and an external Microsoft Excel file.

In the next exercise, we will learn how to control the placement and layout of a TABLE object to fit more easily on a sheet layout.

Fitting a Table on the Page

First, we need to edit the table to fit on the sheet layout:

1. Continue using the 5-3_Using Table Links.dwg file.

2. Select the FURNITURE SCHEDULE table, and select the **Break Table** grip ▼ to drag it UP into the table until the table breaks into two pages and fits on the sheet layout.

FURNITURE SCHEDULE					
Name - (External)	CAPPN	CAPMG	CAPTG	COST	Count
P_2006	2006	STC	MIGRATIONS	$100.00	3
P_2036-WLAM	2036-WLAM	STC	36	$101.00	1
P_287	287	STC	CLUB	$102.00	2
P_421482N	421482N	STC	TRILOGY	$103.00	137
P_462LEAP9L	462LEAP9L	STC	LEAP	$104.00	72
P_50622625	50622625	STC	48/30	$105.00	5
P_730581	730581	STC	18/64/4	$106.00	19
P_851204838	851204838	STC	120/48	$107.00	2
P_853525	853525	STC	35/20	$108.00	1
P_854800	854800	STC	48	$109.00	1
P_85723630	85723630	STC	72/36	$110.00	1
P_85964236	85964236	STC	96/42	$111.00	2
P_9243A	9243A	STC	75/25	$112.00	1
P_9259A	9259A	STC	60/30	$113.00	61
P_9259AR	9259AR	STC	60/30	$114.00	11
P_9283A	9283A	STC	90/30	$115.00	65
P_98013	98013	STC	30/42	$116.00	5
P_98013P3	98013P3	STC	30/42	$117.00	1
P_981830T	981830T	STC	TB/30	$118.00	1
P_981860T	981860T	STC	TB/60	$119.00	3
P_98213	98213	STC	30/65	$120.00	300
P_98213P3	98213P3	STC	30/65	$121.00	91
P_98215	98215	STC	45/65	$122.00	4
P_98215P3	98215P3	STC	45/65	$123.00	31

FURNITURE SCHEDULE					
Name - (External)	CAPPN	CAPMG	CAPTG	COST	Count
P_986831DA15S	986831DA15S	STC	1	$124.00	108
P_99004BEWP	99004BEWP	STC	45/20	$125.00	60
P_99365	99365	STC	BRK	$126.00	118
P_99409	99409	STC	CD	$127.00	68
P_99995	99995	STC	30/45/45/30	$128.00	1
P_9BBL3015	9BBL3015	STC	SB/30	$100.00	129
P_9LF18302F	9LF18302F	STC	LF/30	$101.00	9
P_9LF18305F	9LF18305F	STC	LF/30	$102.00	79
P_9TT6015	9TT6015	STC	60/15	$103.00	2
P_9U224	9U224	STC	BBF	$104.00	3
P_9U947	9U947	STC	FF	$105.00	58
P_9U948	9U948	STC	BBF	$106.00	70
P_CLEJ3660	CLEJ3660	STC	60/36	$107.00	1
P_G23	G23	STC	ESCAPADE	$108.00	4
P_GCB4224	GCB4224	STC	42/24	$109.00	1
P_GCD7236	GCD7236	STC	72/36	$110.00	2
P_GCN7224PKP	GCN7224PKP	STC	72/24	$111.00	2
P_GCTN7236	GCTN7236	STC	72/36	$112.00	1
P_GCTR4848	GCTR4848	STC	48	$113.00	1
P_GCV72	GCV72	STC	SM/72	$114.00	1
P_LSB24KD	LSB24KD	STC	SLU/25	$115.00	128
P_W45BC3672	W45BC3672	STC	BK/5H/36	$116.00	4

Figure 5.28: Breaking the Table into parts

Breaking a Table into Parts

When the TABLE object is larger than expected, it can become necessary to move and separate parts of the table on the sheet layout:

1. Continue using the 5-3_Using Table Links.dwg file.

2. Using the **Properties** dialog, locate the **Table Breaks** section and modify the following settings:

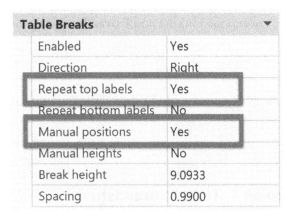

Figure 5.29: Table Properties

3. Select the **Move Table** grip ▢ and move the second page of the table to the other side of the sheet layout.

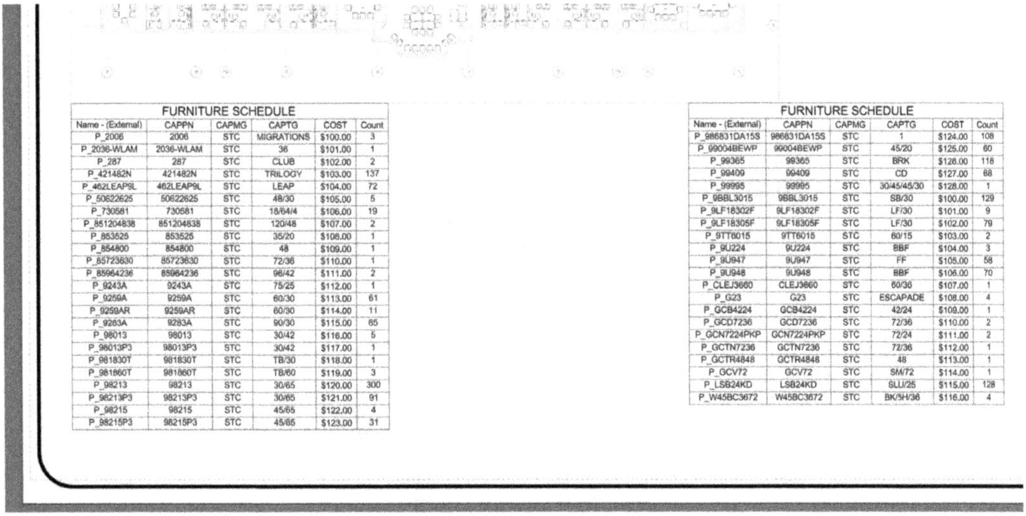

Figure 5.30: Separating Table Pages

Updating a Table Link

You can update the table links during the project using the UPDATE TABLE DATA LINKS command:

1. Continue using the 5-3_Using Table Links.dwg file.

2. Select the FURNITURE SCHEDULE TABLE object.

3. Using the **Insert** ribbon and the **Linking and Extraction** panel, select **Download From Source** command 🔃.

> **Note**
>
> You can also use the *right-click* menu to access **Update Table Data Links** command.

A linked table cell is locked to prevent changes from being made. You can unlock a data link using the UNLOCK command.

Removing a Link to an External Spreadsheet

1. Continue using the 5-3_Using Table Links.dwg file.

2. Using the **Insert** ribbon and the **Linking and Extraction** panel, select the **Data Link** command. 🗒️

3. Using the **Data Link Manager** dialog, select the **Furniture** Excel link and *right-click* to access the **Delete** command.

In the next section, we will learn how to link one TABLE object to another TABLE object to share data between the two tables.

Linking a Table to Another Table

Did you know that you can also link a cell from one table to a cell in another table to share data?

1. Open the 5-3_Using Table Links.dwg file.

2. Using the In-Canvas View Controls, restore the **Custom Model Views | 2-Link Table Cells** named view.

3. Select the **Rebar Total Costs** table by issuing a *left-click* "inside" the E3 cell.

	A	B	C	D	E	F
1			**Rebar Total Costs**			
2	SIZE	DIAMETER	LBS/FT	COST/FT	TOTAL LENGTH	TOTAL COST
3	3	0.375	0.376	0.15		
4	4	0.5	0.668	0.15		

Figure 5.31: Selecting the Table Cell Destination

4. *Right-click* to access the **Insert | Field** command.

5. Using the **Field** dialog, select **Field Name | Formula**.

6. Click the **CELL** button and *left-click* "inside" the **4'-8 1/2"** cell in the **Estimated Quantities** TABLE object.

7. Modify **Format** to **Decimal** and click the **OK** button to close the dialog.

Estimated Quantities			
BAR	**NO.**	**LENGTH**	**TOTAL LENGTH**
301	4	44'-8"	
401	8	4'-4½"	
501	20	4'-8 1/2"	
601	2	14'-7"	
701	6	44'-8"	
	TOTAL		

Rebar Total Costs					
SIZE	**DIAMETER**	**LBS/FT**	**COST/FT**	**TOTAL LENGTH**	**TOTAL COST**
3	0.375	0.376	0.15	56.50	
4	0.5	0.668	0.15		
5	0.625	1.043	0.15		
6	0.75	1.502	0.20		

Figure 5.32: Linking Table Cells 1

Using Table Styles

We all use common tables of information in our drawing files. The key to using the new TABLE object efficiently is to define common Table Styles, which allow us to define and control the format and appearance for quick placement in our drawing files.

Because tables are not yet annotative, you will want to define text sizes for placement on a sheet layout to work for all drawing scales.

Table Settings

First, let's look at how to create a new Table Style:

1. Open the 5-4_Using Table Styles.dwg file.

2. Using the In-Canvas View Controls, restore the **Custom Model Views | 1-Using Table Styles** named view.

3. Using the **Annotate** ribbon and the **Tables** panel, select the "dialog launcher" icon ◥ to open the **New Table Style** dialog.

4. Use the following settings to define a new Table Style:

   A. Use the **Table Start From** button ▦ to choose an existing table in a drawing to use as a "template" for a new table style.

   B. Use the **Table Direction** *drop-down list* to select the direction of the table, either DOWN or UP. This determines whether the Title row is on the TOP or BOTTOM of the table.

C.   This preview window displays the table using the current Table Style settings.

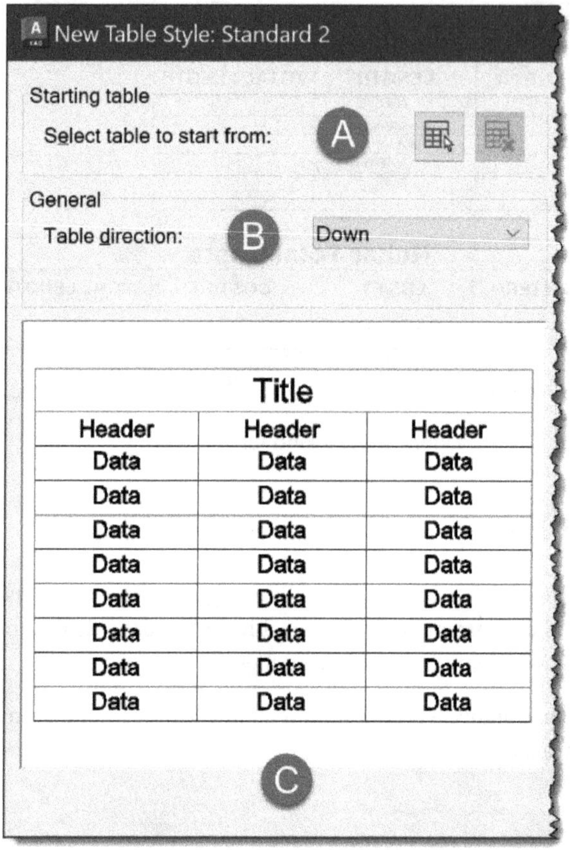

Figure 5.33: Defining the Table Style

Cell Styles

In this section, we will learn how to create Cell Styles to use in our custom tables.

The General Tab

Use this list to review the settings available on the CELL STYLES | GENERAL Tab.

A.   Select the cell style to display the properties of new and existing cell styles.

B.   Define the background color for the selected cell style. The default is **None**.

C.   Define the justification of the selected cell style. The default is **Top Center**.

D.   Define the data type format of the selected cell style.

E.   Define the usage type of the selected cell style, either **Data** or **Label**.

F.   Define the interior margins between the selected cell style data and borders.

G.   Control whether the cell style will automatically merge during the creation of new rows and columns.

H.   Displays a preview of the selected cell style.

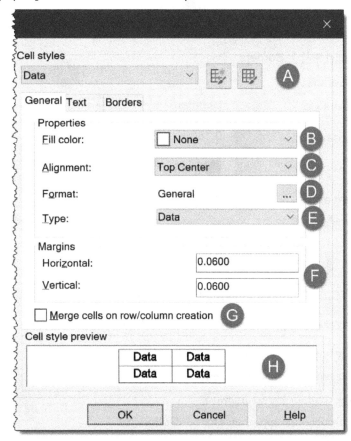

Figure 5.34: Defining a Cell Style – the General Tab

The Text Tab

Use this list to review the settings available on the CELL STYLES | TEXT Tab.

A.   Define the text style used in the cell style.

B.   Define the text height used in the cell style.

C.   Define the text color used in the cell style.

D.   Define the text angle used in the cell style.

E.   Open the **Text Style** dialog to define or modify a text style.

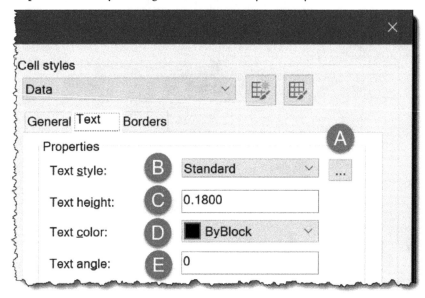

Figure 5.35: Defining a Cell Style – the Text Tab

The Borders Tab

Use this list to review the settings available on the CELL STYLES | BORDERS Tab.

A.   Define the lineweight for border linework.

B.   Define the linetype for border linework.

C.   Define the color for border linework.

D.   Define a Double-line appearance for border linework.

E.   Define the spacing applied to the Double-Line linework.

F.   Apply defined linework to selected border linework.

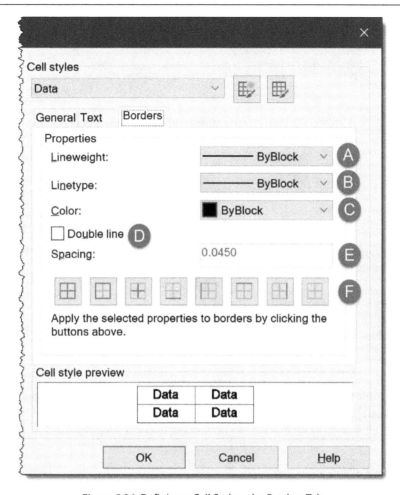

Figure 5.36: Defining a Cell Style – the Borders Tab

Creating a Table Style From an Existing Table

In this exercise, we will learn how to create a new Table Style from an existing TABLE object.

1. Continue using the 5-4_Using Table Styles.dwg file.

2. Using the In-Canvas View Controls, restore the **Custom Model Views | 2-From Existing Table** named view.

3. Select the **SCHEDULE** table and *right-click* to access the **Table Style | Save As New Table Style** command.

4. Key in MY SCHEDULE for the style name and click the **OK** button to save and close the dialog.

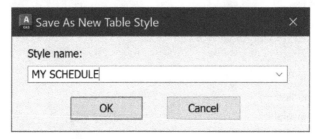

Figure 5.37: Saving as New Table Style

5. Using the **Annotate** ribbon and the **Tables** panel, set the current table style to **MY SCHEDULE**.

6. Select the **Table** command and place the new table using the new style, **MY SCHEDULE**, and place the table in the drawing file.

Sharing Table Styles

You can share Table Styles between files using the following methods.

Method 1

Using the Clipboard to COPY and PASTE a table from one drawing to another copies the TABLE object and all its settings, such as table and cell styles.

Method 2

Using DesignCenter allows you to copy drawing styles and graphics into other drawing files, such as Table Styles.

ADCENTER	Command Locations
Ribbon	View	Palettes	DesignCenter
Command Line	ADCENTER (AD)

In this exercise, we will use the DesignCenter dialog to transfer styles from another drawing file into the current drawing file.

1. Continue using the 5-4_Using Table Styles.dwg file.

2. Using the **View** ribbon and the **Palettes** panel, select the **DesignCenter** command.

3. Using the **DesignCenter** palette, select the **Folders** tab, navigate to the following location, and select the Table Styles file shown here:

    ...ACAD_TipsTechniques\Exercise_Files\CompanyXYZ\standards\
    XYZ_Standard Table Styles.dwg

> **Note**
>
> If you *right-click* on the Standards file, you can select **Add to Favorites** to allow for quicker access to your standard files in the future.
>
> You can also *right-click* on the Standards folder and select **Set as Home** to gain quicker access for future navigation.

4. Use the **+** symbol to expand the contents of the `Standard Table Styles.dwg` file.

5. Select the **Table Styles** content, and *left-click and drag* any table style into your current drawing file.

> **Note**
>
> You can *right-click* and *drag and drop* any data from the DesignCenter into your current drawing file to get the following options:

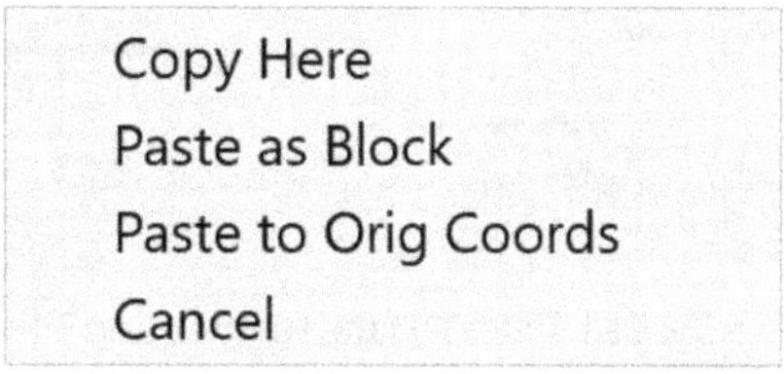

Copy Here
Paste as Block
Paste to Orig Coords
Cancel

Figure 5.38: DesignCenter drag and drop options

Using Cell Styles

When working with TABLE objects, it is helpful to define standard table cell styles that can be reused in multiple standard Table Styles.

Creating Cell Styles from Scratch

Cell styles can be created from scratch or from existing table cells for commonly used cell formats and appearances:

1. Continue using the `5-4_Using Table Styles.dwg` file.

2. Using the In-Canvas View Controls, restore the **Custom Model Views | 3-From Scratch** named view.

3. Using the **Annotate** ribbon and the **Tables** panel, select the **Table Style** dialog launcher ⬐.

4. Select the **MY SCHEDULE** table style and click the **Modify** button.

5. Using the **Modify Table Style** dialog, locate the **Cell styles** section to define the **Data**, **Header**, and **Title** cell style properties.

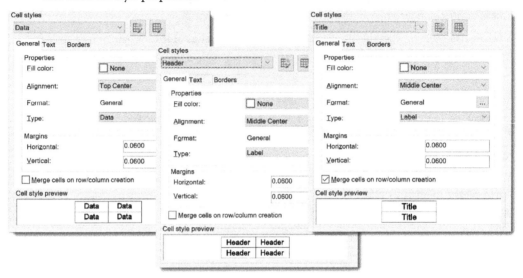

Figure 5.39: Cell Styles – Data, Header, and Title

6. Using the **Cell styles** *drop-down list*, select the **Create new cell style…** command.

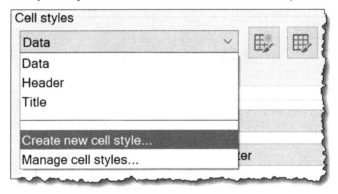

Figure 5.40: Creating a New Cell Style

7. Using the **Create New Cell Style** dialog, key in LEGEND TITLE as the new **Cell Style** name, and modify **Start With:** to use the existing **Title** style.

8. Click **Continue** to return to the **Modify Table Style** dialog.

9. Using the **Modify Table Style | Cell Styles | Properties** sections, modify the following settings for the **General** and **Text** tabs:

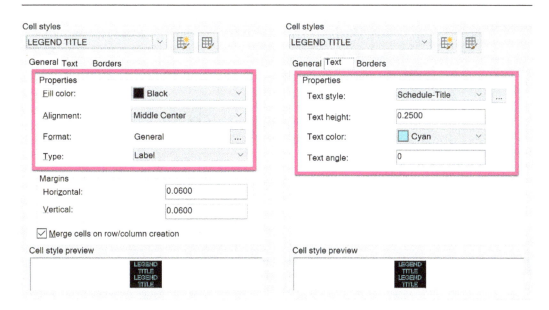

Figure 5.41: Cell Style General and Text Properties

10. Click **OK** to close the dialog and save these changes, and click **Close** to close the **Table Style** dialog.

Creating Cell Styles from an Existing Cell

You can also create a cell style from an existing table in the drawing:

1. Continue using the 5-4_Using Table Styles.dwg file.

2. Using the In-Canvas View Controls, restore the **Custom Model Views | 4-From Existing Cell** named view.

3. Select the existing **LEGEND** TABLE object | **Title** cell, and *right-click* to access the **Cell Style | Save as New Cell Style...** command.

4. Key in the name Title-FILLED for the style name and click **OK** to save the new cell style.

5. Using the existing **LEGEND** TABLE object | **TOTAL** cell, and *right-click* to access the **Cell Style | Save as New Cell Style...** command.

6. Key in the name Total-FILLED for the style name and click **OK** to save the new cell style.

7. Click the **OK** button to save and close the dialog.

    Next, we will place a TABLE object and use the new Cell Styles.

8. Using the **Annotate** ribbon and the **Tables** panel, set the current Table Style to **MY SCHEDULE**.

9. Select the **Table** command and click **OK** to close the dialog.

10. *Left-click* below the existing **LEGEND** TABLE object to place the new table in the drawing.

11. Use the *Esc* key to complete the command.

12. Select the **SCHEDULE** TABLE object | **Title** cell and, using the **Properties** dialog, make the following change:

Cell Style: LEGEND TITLE

> **Note**
>
> The only Cell Styles available are those that were created for the MY SCHEDULE table style. The filled cell styles were generated for the STANDARD table style.

Let's look at how to create a custom cell style with a filled background:

1. Continue using the 5-4_Using Table Styles.dwg file.

2. Using the **Annotate** ribbon and the **Tables** panel, set the current Table Style to **STANDARD**.

3. Select the **Table** command and click **OK** to close the dialog.

4. *Left-click* to the RIGHT of the existing LEGEND table to place the new table in the drawing.

5. Select the **SCHEDULE** TABLE object | **Title** cell, and, using the **Properties** dialog, make the following change:

Cell Style: Title-FILLED

6. Select the TOTAL cell and *right-click* to access the **Cell Style | Total-FILLED**.

> **Note**
>
> You can use the MATCH CELL command to match cell styles within the same table. This command copies all the properties of the cell except the cell type: TEXT or BLOCK.
>
> MATCH CELL can only be used within the same TABLE object, and it only works if the cell type is the same (i.e., text, decimal, whole number).

Using Tables and Tool Palettes

Using tables on standard Tool Palettes provides easy access to common schedules.

Placing Tables on a Tool Palette

Put your common TABLE objects on a Tool Palette for easy access in the next drawing. Once a standard table has been completed, you can place it on a Tool Palette using the following steps:

1. Open the 5-1_Working with Tables.dwg file.

2. Using the In-Canvas View Controls, restore the **Custom Model Views | 4-Add Table to Palette** named view.

3. Select the existing **SCHEDULE 1** table from the drawing.

4. Use the **View** ribbon and the **Palettes** panel to display your Tool Palettes.

TOOLPALETTES OPEN	Command Locations
Ribbon	View	Palettes	Tool Palettes
Shortcuts	Ctrl + 3 (toggles OPEN and CLOSE)
Command Line	TOOLPALETTES (TP) to OPEN, TOOLPALETTESCLOSE to CLOSE

> **Note**
>
> A blank Tool Palette is provided for you to practice this exercise in the course and can be found at this location:
>
> ...\ACAD_TipsTechniques\Exercise_Files\CompanyXYZ\toolPalette

1. *Right-click* on the mouse to *drag and drop* the **SCHEDULE 1** table onto the **ACAD_TipsTechniques** Tool Palette.

2. Place the new schedule from the Tool Palette into the drawing.

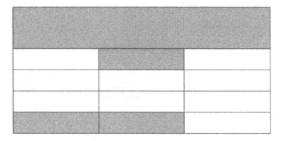

Figure 5.42: Table Object from the Tool Palette

By default, you will notice that all Title, Header, and Data values are removed from the table when placed from a saved style from a Tool Palette, and the size is not retained using this method.

To retain the cell values, we need to modify the properties of the table on the Tool Palette.

3. Using the Tool Palette, *right-click* on the new table tool to access the **Properties** command.

4. Scroll down to locate the **Insert Options** and modify the **Label Cells** option to YES. Again, this doesn't correct the size issues when placing the table from a Tool Palette.

A better option is to make a block of the table and insert the block from the Tool Palette.

Reusing Tables with Data

If you need the data in a table to remain in the table for reuse, make the table into a block and then place the block containing the table on the Tool Palette. Set up the "block table" to EXPLODE on INSERT to convert the block back to a table for reuse:

1. Continue using the 5-1_Working with Tables.dwg file.

2. Using the In-Canvas View Controls, restore the **Custom Model Views | 5-Using Tables as Blocks** named view.

3. Select the **SCHEDULE 2** TABLE object. This table has been made into a block object so that it can retain the table values when placed on a Tool Palette.

4. *Right-click* and *drag and drop* the TABLE BLOCK object onto the **ACAD_TipsTechniques** Tool Palette.

   Next, we want the block to explode automatically when inserted, so the resulting object is a TABLE, not a BLOCK object.

5. *Right-click* on the new TABLE BLOCK object on the Tool Palette to access the **Properties** command.

6. Using the **Tool Properties** dialog, locate the **Insert** panel and modify the **EXPLODE** setting to **YES**.

7. Click **OK** to close the **Tool Properties** dialog.

8. Using the **ACAD_TipsTechniques** Tool Palette, *left-click* on the **SCHEDULE** TABLE object and place the table in the drawing.

9. Select the new TABLE object to verify that the block exploded on insert and that the new object is a TABLE and not a BLOCK.

SCHEDULE 2		
NAME	SIZE	QUANTITY
D101	3' x 7'	2
D102	3'6 x 7'	3
D103	4' x 7'	1
D104	4' x 7'6"	5
TOTALS		11

Figure 5.43: Table BLOCK from the Tool Palette

The problem with this method is you retain all the data values unless you clear them in the block object, and it doesn't use the INSERT TABLE command, so you lose some of the INSERT TABLE command options.

Bonus Table Commands

In this section, we will review some quick tips that can make working with tables a little easier.

How do you redefine a minimum row height?

You can define a minimum row height for when you select the TABLE object and set the cell height to a very small distance. This will reset all rows in the table to their minimum value based on the individual cell text height settings:

1. Open the 5-5_Bonus TABLE Commands.dwg file.

2. Using the In-Canvas View Controls, restore the **Custom Model Views | 1-Minimum Row Height** named view.

3. Select the TABLE object and *left-click* inside cell **A1**. Hold down the *Shift* key and *left-click* inside cell **C7** to select all the cells in the table.

4. Using the **Properties** dialog, locate the **Cell Height** property and modify the height to .001.

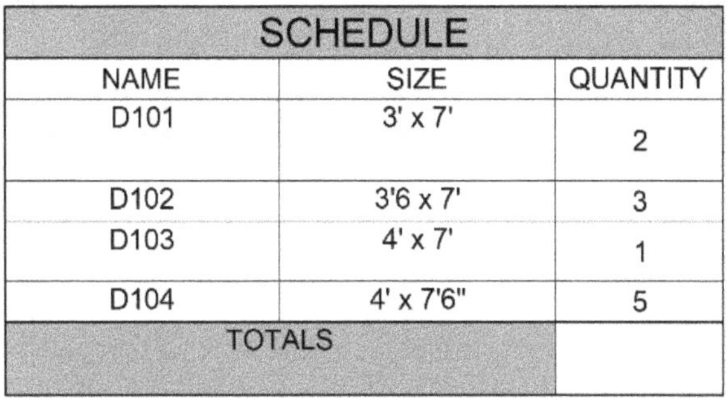

SCHEDULE		
NAME	SIZE	QUANTITY
D101	3' x 7'	2
D102	3'6 x 7'	3
D103	4' x 7'	1
D104	4' x 7'6"	5
TOTALS		

Figure 5.44: Cell Height Initial Graphics

The resulting graphics are easily adjusted with minimal row heights.

SCHEDULE		
NAME	SIZE	QUANTITY
D101	3' x 7'	2
D102	3'6 x 7'	3
D103	4' x 7'	1
D104	4' x 7'6"	5
TOTALS		

Figure 5.45: Cell Height Results

How do you save a new table style over an existing table style?

To overwrite an existing table style, you must use the SAVE AS NEW TABLE STYLE command, which is only found on the *right-click* menu when a table is selected:

1. Continue using the 5-5_Bonus TABLE Commands.dwg file.

2. Using the In-Canvas View Controls, restore the **Custom Model Views | 2-Overwrite Table Style** named view.

3. Select the modified table, SCHEDULE, *right-click* to access the **Table Style | Save as New Table Style** command, and key in a duplicate name. When prompted to redefine the existing table style, verify that you want to overwrite the existing style.

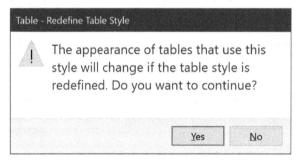

Figure 5.46: Redefining a Table Style

How do you turn off the display of the "hidden" lines in a table?

When viewing a TABLE object, the hidden lines, or lines that have no border defined, appear as thin gray lines. These lines are automatically placed on the DEFPOINTS layer.

1. Continue using the 5-5_Bonus TABLE Commands.dwg file.

2. Using the In-Canvas View Controls, restore the **Custom Model Views | 3-Hidden Table Lines** named view.

3. Using the **Home** ribbon and the **Layers** panel, **FREEZE** the **Defpoints** layer.

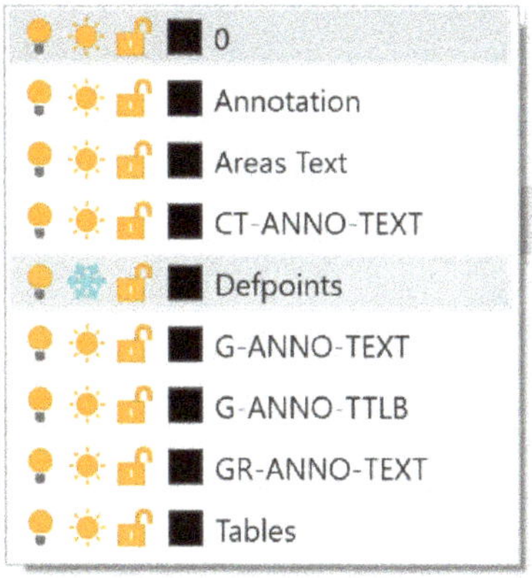

Figure 5.47: Table "hidden" Linework

Here, you can see the initial hidden lines in a TABLE object:

STANDARD TABLE		

Figure 5.48: Initial Table Hidden Lines

Here, the initial hidden lines in the table are turned off using the DEFPOINTS layer:

STANDARD TABLE

Figure 5.49: Table Hidden Lines Results

Summary

In this chapter, we learned how to work more efficiently with the TABLE object using keyboard shortcuts for navigation and GRIPS for quick modifications. We learned how to control the values in our TABLE objects using formulas, equations, and cell locking. After that, we learned how to make smarter tables using FIELDS to capture and automate values from drawing objects, external data, and other table values. Once all these skills were mastered, we learned how to create Table and Cell Styles to better control the standardization of TABLE objects in our drawings, including how to put our common TABLE objects on a Tool Palette for easy recall.

In the next chapter, we will examine how to improve our use of blocks, including attributes, custom basepoints, block libraries, and groups.

6

Discover More About Blocks

In this chapter, you will learn how to use blocks to their fullest by adding the ability to automate daily tasks when using blocks. You will learn how to control blocks when inserting them into your drawings with or without attributes. You will also learn to create easy-to-use block libraries, and work with some lesser-known block commands that many of you will find handy to know.

In this chapter, we'll cover the following topics:

- Working efficiently with blocks

- Working with attributes

- Bonus block commands

By the end of this chapter, you will be able to use and share your BLOCK objects more efficiently and take advantage of the new smart blocks technology and data associated with your blocks.

Working efficiently with blocks

In this section, we will discuss how normal blocks work in AutoCAD and why we should use common settings such as Layer 0, ByLayer, and ByBlock. We will also learn how to create cleaner blocks and non-explodable blocks and how to use on-the-fly blocks when they are appropriate. We will investigate how to modify the default insert options when placing blocks to take advantage of alternative basepoints, which can make your blocks more user-friendly.

Why Layer 0

I am always amazed at how often I get the question, "Why do we create blocks on layer 0?" So, let me clarify this for you.

Layer 0 is our "default" or "generic" layer in AutoCAD. When a block's graphics are created on Layer 0, the block will automatically inherit the Active Layers properties when placed in the drawing, allowing us to use a single block for multiple purposes. When the graphics in the block have their properties defined as ByLayer for Color, Linetype, Lineweight, and so on, they inherit the properties from the Active Layer when placed. This also applies to References (Xrefs) that contain objects on Layer 0.

If you place graphics in a block that is not on Layer 0, those graphics will maintain their defined properties for Layer, Color, Linetype, and Lineweight, regardless of what layer the block is placed on.

BYLAYER versus BYBLOCK

Another question that is often asked during a training session is, what is the difference between BYLAYER and BYBLOCK? Initially, they may appear to work the same. However, they both allow you to insert blocks using the active property settings and layers when needed. Let me explain further.

BYLAYER

When your block's graphics are created using BYLAYER for Color, Lineweight, or Linetype, the inserted block's appearance will always defer to the active layer's properties on which it is placed.

BYBLOCK

When the graphics of your block are created using BYBLOCK for color, lineweight, or linetype, the appearance of the inserted block will always defer to the active properties when the block is placed and will ignore the layer's properties that the block is placed on. When the active property value is set to a specific color, linetype, or lineweight, the inserted block will inherit these active property settings. This is beneficial when a block serves more than one purpose and needs to be displayed using various properties such as existing, proposed, and demolition representations.

Many users do not take advantage of the BYBLOCK functionality in AutoCAD. Hopefully, this will inspire you to use more of the BYBLOCK capabilities.

In this exercise, we will learn how to use AutoCAD's BYLAYER and BYBLOCK features:

1. Open the `6-1_Working with Blocks.dwg` file.
2. Using the In-Canvas View Controls, restore the **Custom Model Views | 1-Using ByLayer or ByBlock** named view.

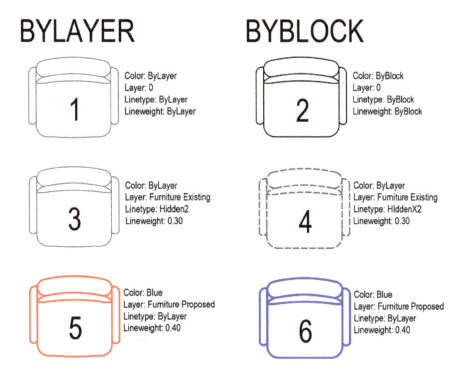

Figure 6.1: ByLayer versus ByBlock

Here, you can see the differences between placing a block that is made using BYLAYER and another using BYBLOCK. The BYLAYER block will always adhere to the properties defined by the layer, while the BYBLOCK block allows you to "tweak" the common properties when needed.

3. Using the **Home** ribbon and the **Layers** panel, change the active layer to **1-Furniture**.

4. Using the **Home** ribbon and the **Properties** panel, change the following settings.

Figure 6.2: Home | Properties settings

5. Select the **CHAIR BYBLOCK** block (2) and *right-click* to access the **Add Selected** command.

Figure 6.3: ByBlock using Add Selected

6. Notice that using this method for inserting the block will retain all property settings from the selected block and ignore the active properties.

7. Use the *Esc* key to cancel the current command and issue *Ctrl + Z* to **UNDO** the previous insertion.

8. Using the **Insert** ribbon and the **Block** panel, select the **Insert Block** command and select the **CHAIR BYBLOCK** block.

9. *Left-click* to add this block to the third column in the drawing view.

Figure 6.4: ByBlock using Insert Block 2

10. Using the **Home** ribbon and the **Properties** panel, change the following settings.

Figure 6.5: Ribbon Properties settings

11. Use the INSERT BLOCK command to place another **CHAIR BYBLOCK** block.

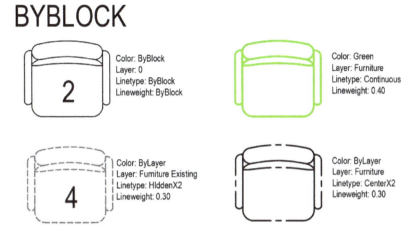

Figure 6.6: ByBlock using Insert Block 2

12. Review your block usage within your organization to discover whether you have circumstances where BYBLOCK might be more flexible to use than BYLAYER.

Make Non-Explodable Blocks

We have all opened that drawing where a block has exploded for some reason, and we have to resolve or fix that issue, right? Let's make it harder to explode our blocks so users will be less likely to do so:

1. Continue using the 6-1_Working with Blocks.dwg file.

2. Using the In-Canvas View Controls, restore the **Custom Model Views | 2-Non-Explodeable Blocks** named view.

3. Select the existing block and *right-click* to access the **Block Editor** command.

4. Using the **Block Editor**, use the **Properties** dialog to locate the Block panel and modify the **Allow Exploding** setting to **No**.

5. Using the **Block Editor** ribbon and the **Close** panel, select the **Close Block Editor** command.

6. Using the **Block – Changes** dialog, select the **Save the Changes** option.

7. Using the **Home** ribbon and the **Modify** panel, select the **Explode** command and select the existing block.

You can no longer explode this block. Be sure to make blocks non-exploding where necessary.

In the next exercise, we will learn how to modify the block's insertion point on the fly when inserting a block.

On-the-Fly Basepoints

By default, when placing a block in a file, the insertion point is controlled by the 0,0 location defined for that block's graphics. However, it would be helpful to provide more than one insertion point for more flexibility on some blocks. You can temporarily alter the BASEPOINT POINT or INSERTION POINT during the INSERT command. Let me show you how:

1. Continue using the `6-1_Working with Blocks.dwg` file.
2. Using the In-Canvas View Controls, restore the **Custom Model Views | 3-Using Basepoints** named view.
3. Select the existing CHAIR block object (CHAIR 1).
4. *Right-click* to access the **Add Selected** command.
5. The CHAIR block is automatically selected.
6. The CHAIR block is attached to the cursor based on the **0,0** point defined in the block.
7. Using the Command Line, key in B to access the **Basepoint** command option.

> **Note**
>
> You can also select the BASEPOINT command option in the Command Line.

8. Select the FRONT MIDPOINT of the seat to redefine a temporary insertion point.
9. Place the CHAIR block at the TOP of the TABLE furniture object.

The next exercise will demonstrate how to insert blocks at non-existent points.

Use Temporary Tracking with Insert Block

We can also place the BASEPOINT referenced off the front of the CHAIR block using a TEMPORARY TRACKING point to place the CHAIR block under the edge of the TABLE object:

1. Continue using the `6-1_Working with Blocks.dwg` file.
2. Select the existing **CHAIR** block object (CHAIR 1).
3. *Right-click* to access the **ADD SELECTED** command.
4. The **CHAIR** block is attached to the cursor based on the **0,0** point defined in the block.
5. Using the Command Line, key in B to access the **Basepoint** command option.
6. Using the Command Line, key in TT to define a TEMPORARY TRACKING point at the FRONT MIDPOINT of the **CHAIR** object as a "temporary location" and *left-click* to accept this temporary point.

7. *Left-click and drag* the cursor DOWN, key in the distance of 9 inches (9 units), and *left-click* to accept this new insertion point.

8. Use the **CENTER** OSNAP to snap to the center of the **TABLE**.

9. Using the Command Line, use the *Enter* key to accept the X and Y scale as 1, and key in the 180 angle to place the chair.

> **Note**
>
> Use the DRAWORDER command to put the CHAIR object behind the TABLE object containing the WIPEOUT object.

> **Note**
>
> You can also swing the CHAIR block into position using POLAR TRACKING lines to get the 180-degree angle.

In the next exercise, we will learn how to define "alternate" basepoints in the BLOCK itself using POINT parameters.

Defining Alternate Basepoints

You can build alternate BLOCK insertion points inside the block using POINT parameters in a dynamic block definition:

1. Continue using the 6-1_Working with Blocks.dwg file.

2. Using the In-Canvas View Controls, restore the **Custom Model Views | 4-Alternate Basepoints** named view.

3. Select the existing **CHAIR** block object (CHAIR2) and *right-click* to access the **Block Editor** command.

4. Using the **Block Authoring** palette, select the **Parameters** tab and select the **Point** parameter.

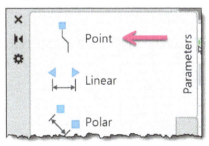

Figure 6.7: Using a POINT parameter

5. *Left-click* to place a POINT parameter at the four additional basepoint locations shown in the following figure.

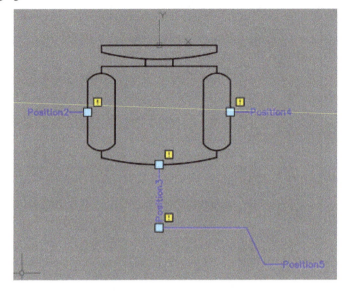

Figure 6.8: Placing POINT Parameters

Note

You can ignore the exclamation points for now; this will be discussed in further detail in *Chapter 7*. An exclamation point normally indicates that the parameter is not completely defined.

It is helpful to assign logical names and descriptions to these POINT parameters.

6. Select the **Position2** POINT parameter.

7. Using the Block Editor and the **Properties** dialog, locate the **Property Labels** setting and change the **Property Name** setting to **Right Arm** (when you are sitting in the chair).

Note

You can also use a *right-click* to access the RENAME PARAMETER command.

---

8. Continue renaming the following POINT parameters:

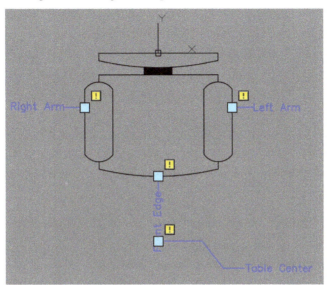

Figure 6.9: POINT parameter names

You can control the basepoint insertion order using the BCYCLEORDER command when using the **Block Editor**:

1. Using the Block Editor, key in the BCYCLEORDER command.

2. Using the **Insert Cycling Order** dialog, use the **Move Up** and **Move Down** buttons to change the order of the insertion points.

3. You can also use the **Cycling** button to disable the ability to use any of the POINT parameters as an alternate insertion point.

> **Note**
>
> I'm sorry that the Parameter Names are not displayed in this dialog. This worked in previous versions and hopefully, Autodesk will fix this soon!

Now, we can test the new block functionality without exiting the Block Editor.

4. Using the **Block Editor** tab, select the **Test Block** command.

5. Using the **Insert** ribbon and the **Block** panel, select the **Insert Block** command 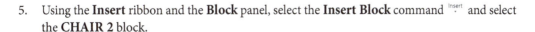 and select the **CHAIR 2** block.

6. Use the *Ctrl* key to toggle between the new "alternate" insertion points.

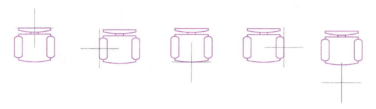

Figure 6.10: Alternate block basepoints

7. Use the **Close Test Block** command to end the testing environment and use the **Close Block Editor** command to close and save the **CHAIR2** block.

> **Note**
>
> The "alternate" insertion points work with the ADD SELECTED command, but you need to move the cursor slightly to see the updated basepoint locations.

In the next section, we will learn how to gain quicker access to our commonly used and favorite blocks in a drawing.

Quickly Access Recent and Frequently Used Blocks

In this section, we will use the new **Blocks** dialog provided in 2024 and later versions of AutoCAD. The Recent tab displays the previously inserted blocks regardless of the current drawing and will persist between drawings and AutoCAD sessions.

This exercise will investigate how to use the Recent tab to access our previously used blocks:

1. Continue using the 6-1_Working with Blocks.dwg file.

2. Using the In-Canvas View Controls, restore the **Custom Model Views | 5-Recently Used Blocks** named view.

3. Using the **Insert** ribbon and the **Block** panel, select the **Insert Block** *drop-down list* to access the **Recent Blocks** command to open the **Blocks** palette.

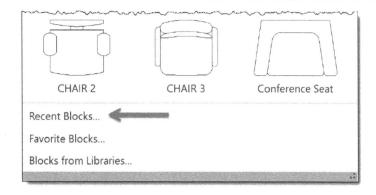

Figure 6.11: Using Recent Blocks

4. Using the **Blocks** palette, select the **Recent** tab.

5. Use the TOP portion of this palette to review your recently used blocks, and make note of the **Filter** option to easily control how many blocks are displayed.

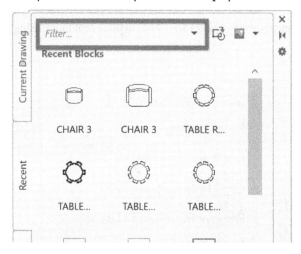

Figure 6.12: Recent Blocks filter

> **Note**
> You can insert blocks from any of the Block palette tabs using *drag-and-drop* or by issuing a *left-click* to insert the block with OSNAP precision.

6. Be sure to modify the thumbnail display to your preferred display. Long block names can be more easily found using the **List** option if the preview isn't needed.

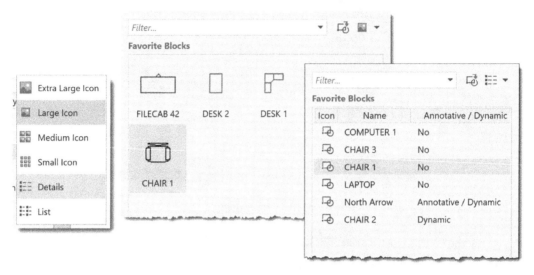

Figure 6.13: Block thumbnail controls

7. Using the **Favorites** tab, *right-click* on any block to place that block using the insertion options shown here.

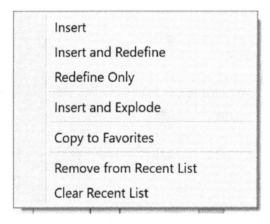

Figure 6.14: Insertion options

> **Note**
>
> You can **Insert and Redefine** a block in the current drawing, **Redefine Only**, or **Insert and Explode** the block immediately on insert.

8. Use the BOTTOM portion of this palette to refine the **Insert Block** command options for placing the block again.

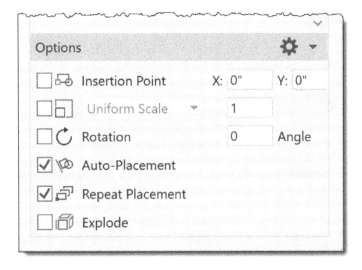

Figure 6.15: Continuous Insert Block command options

> **Note**
>
> Be sure to toggle on the **Repeat Placement** option to simplify multiple block insertions.

This section provides a much easier way to place a block without the need to issue responses to all the command steps required using the normal INSERT BLOCK command. After all, how often do you change the scale of a block to anything other than 1?

In the next section, we will learn how to set up our "favorite" blocks for quicker access.

Favorite Blocks

AutoCAD 2024 and later provides the new BLOCKS palette, which includes the FAVORITES tab. This tab allows you to save your commonly used blocks, providing quicker access. In this exercise, we will learn how to use the FAVORITES tab:

1. Continue using the 6-1_Working with Blocks.dwg file.
2. Using the In-Canvas View Controls, restore the **Named View | 6-Favorite Blocks** named view.
3. Using the **View** ribbon, verify that you have the **Blocks** palette open.
4. Using the Blocks palette, select the **Current Drawing** tab and define the filter as CHAIR* to display only the **CHAIR** blocks.
5. *Right-click* on the CHAIR2 block and select the **Copy to Favorites** command.

6. Clear the filter using the **X** icon, locate the **FILECAB 42** block, and *right-click* to access **Copy to Favorites**.

7. Locate the **TABLE CHAIR-Single** block and, again, *right-click* to **Copy to Favorites**.

8. Using the Blocks palette, select the **Favorites** tab. Here, you can access the blocks you use most often and have them easily available.

9. If you need to remove or clear your **Favorites** tab, issue a *right-click* anywhere in the **Favorites** section and select the **Remove from Favorites List** or **Clear Favorites List** commands.

However, what if you have hundreds of blocks that you use on a regular basis? In the next exercise, we will learn how to set up an entire folder of FAVORITES blocks:

1. Place the cursor anywhere in the view window and *right-click* to access the **Options** command.

2. Using the **Options** dialog, select the **Files** tab and expand the **Blocks Sync Folder Location** section.

3. Here, you can **Browse** and navigate to any folder within your organization or your own personal folder structure. I have delivered a folder with sample favorites using electrical block examples.

4. Click the **Browse** button and navigate to the ...\ACAD_TipsTechniques\Exercise_ Files\CompanyXYZ\Block Libraries\Lighting folder.

In the next section, we will learn how to work more efficiently with attributes in our blocks.

Working with Attributes

This section will look at how to do more with the Attributes in your blocks. Assuming you all know how to use ATTRIBUTE objects in a block, we will look at how to use Multi-line Attributes, control the order of those Attributes, and quickly edit an ATTRIBUTE. We will also learn how to use Annotative Attributes, sync changes, globally edit Annotative scales, and scale locations. How can you keep that data when you need to explode that block once your block attributes are populated with data?

Multi-Line Attributes

To improve block functionality, you can create blocks with attributes that contain multi-line strings of text that use UNDERLINE, OVERLINE, and FIELDS:

1. Open the 6-2_Working with Attributes.dwg file.

2. Using the In-Canvas View Controls, restore the **Custom Model Views | 1-Multi-line Attributes** named view.

3. Using the **Home** ribbon and the **Blocks** panel, select the **Blocks** expanded panel to access the **Define Attributes** command. ✎

4. Define the following attribute values as displayed here.

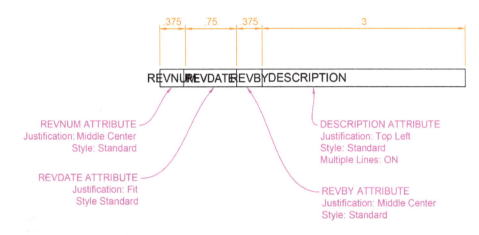

Figure 6.16: Defining a Multi-line Attribute

> **Note**
>
> Use a FIT TEXT object for the DATE attribute to allow larger characters to fit as needed. The FIT TEXT attribute allows us to "squish" text into the defined linework.

5. Using all the linework and attributes, create a new **RevisionBlock** block.

6. *Double-click* on the **BLOCK | RevisionBlock** to edit the attributes.

7. Using the **Enhanced Attribute Editor** dialog, select the **Description** attribute and click the ... button to edit the attribute value.

8. Replace the default value with the REVISION DESCRIPTION REQUIRING MORE THAN ONE LINE OF TEXT text.

9. Use the **Paragraph Width** icon ◂▸ to define the paragraph width of the **Description** attribute.

Next, we will edit the existing block to demonstrate how to control the order of the attributes during edits.

Controlling the Order of Block Attributes

Once the block has been made, you may need to change the order of the block attributes for easier attribute editing. Two methods are available for controlling this order.

Method 1 – Using BATTMAN

Using this method, we will use the BATTMAN command to change the order of the block attributes.

1. Continue using the 6-2_Working with Attributes.dwg file.

2. Using the Command Line, key in the **Battman** command.

3. Using the **Block Attribute Manager** dialog, click the **Select Block** button 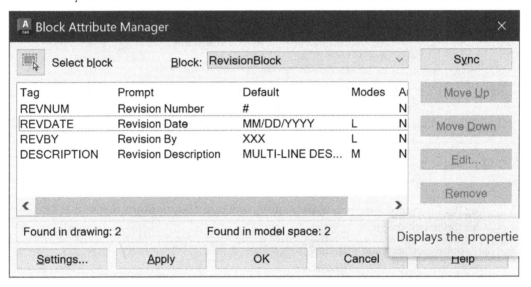 and select the **RevisionBlock** block.

| # | MM/DD/YYYY | XXX | MULTI-LINE DESCRIPTION |

Figure 6.17: Block Attributes

4. Using the **Block Attribute Manager** dialog, use the **Move Up** and **Move Down** buttons to modify the order of the attributes to match the order of the attributes in the block, as shown here.

Figure 6.18: Method 1 – Controlling the Attribute Order

5. Click **OK** to save these changes and close the dialog.
6. *Double-click* on the block to test the changes in the **Enhanced Attribute Editor** dialog.

Method 2 – Using BATTORDER

The BATTORDER command can be used when editing the block using the Block Editor:

1. Continue using the 6-2_Working with Attributes.dwg file.
2. Select the **RevisionBlock** block and *right-click* to access the **Block Editor** command.
3. Inside the **Block Editor** command use the Command Line to key in the **Battorder** command.

4. Using the **Attribute Order** dialog, use the **Move Up** and **Move Down** buttons to modify the order of the attributes to match the order of the attributes in the block.

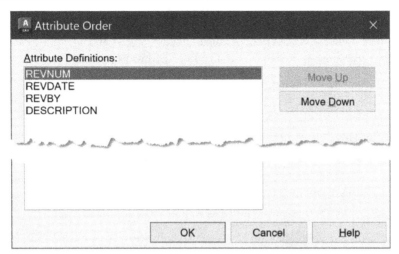

Figure 6.19: Method 2 – Controlling the Attribute Order

5. Click **OK** to save these changes and close the dialog.

Next, we will learn how to use keyboard shortcuts to work more easily with the Block Editor and reduce the number of clicks it takes to close and save block changes.

Quick Exit the Block Editor

You can quickly close the Block Editor using the following shortcut keys:

1. Continue using the `6-2_Working with Attributes.dwg` file.

2. Inside the **Block Editor**, key in *B + C + Space* or *B + C + Enter* to close the **Block Editor** and save the changes.

3. Use the *B + C + Space + Space* shortcut keys to automatically save the changes made using the Block Editor.

> **Note**
>
> When needed, use a dynamic block to easily stretch the linework to accommodate the new multi-line attribute value. *Chapter 8* covers how to work with dynamic blocks.

In the next exercise, we will work with Annotative blocks to improve your workflow.

Annotative Attribute Blocks

The Block Attribute Manager allows you to edit and manage attribute definitions in BLOCK objects. For this example, I want to focus on how you can make your blocks ANNOTATIVE. The best option is to make the block itself ANNOTATIVE, not the individual objects and attribute definitions. This will keep the block's components together as it shrinks and grows as needed.

First, let me demonstrate why you should create your blocks this way:

1. Continue using the `6-2_Working with Attributes.dwg` file.

2. Using the In-Canvas View Controls, restore the **Custom Model Views | 2-Annotative Blocks** named view.

    Here, you have two blocks that appear identical but are very different in their functionality. First, the ANNO_SECTION_1 block is defined as Annotative at the block level, and the ANNO_SECTION-BAD block is defined as Annotative at the object level.

3. Using the Status bar, verify that you have enabled both **Show Annotation Objects** and **Add Scale to Annotation Objects**.

Figure 6.20: Annotation scale

4. Switch the Annotation scale from **1:1** to **1:2** to visually change the existing blocks.

Figure 6.21: Annotation scale example

Notice that the graphics do not scale as needed when the objects are defined as Annotative on the individual objects. The annotation scale works best when all the objects are defined as non-annotative, and the block is assigned as Annotative.

This is easily fixed using the following steps.

5. Continue using the `6-2_Working with Attributes.dwg` file.

6. Switch the Annotation scale from **1:2** back to **1:1** to visually change the existing blocks.

7. Select the ANNO_SECTION-BAD block and *right-click* to access the **Block Editor** command.

8. Select both Attribute objects (text objects) and, using the **Properties** dialog, modify the **Text | Annotative** setting to **NO**.

9. Use the *Esc* key to clear the selection set.

10. Using the **Properties** dialog, modify the **Block: Annotative** setting to **YES**.

11. Using the **Block Editor** ribbon, select **Close Block Editor**, or use the previously mentioned *C + B + Space + Space* shortcut to close and save the **Block Editor** changes.

12. Switch the Annotation scale from **1:1** to **1:2** to visually check our changes to the existing blocks. Both blocks should scale identically after these changes.

This is a good example of how to make your annotative blocks work more effectively.

In the next example, I want to demonstrate how to use the ATTRIBUTE SYNC command to update the blocks in your drawing file after edits have been made using the Block Editor.

Sync Attribute Changes

When editing blocks that contain attributes, the graphics can become "scrambled" or not appear with the changes as intended when the block is already placed in the drawing file. This can be corrected using the ATTRIBUTE SYNC command to avoid having to re-insert the block.

ATTSYNC	**Command Locations**
Ribbon	Home | Block | Synchronize Attributes
	Insert | Block Definition | Synchronize
Command Line	ATTSYNC (ATTS)

In this example, we will learn how to use the Synchronize Attributes to update a modified block with attributes:

1. Continue using the `6-2_Working with Attributes.dwg` file.

2. Using the In-Canvas View Controls, restore the **Custom Model Views | 3-Sync Attributes** named view.

3. Select the **ANNO_SECTION_2** block and *right-click* to access the **Block Editor** command.

4. Select both text ATTRIBUTE objects and, using the **Properties** dialog, modify the **Text Style** setting from **ROMANS** to **Standard**.

5.   Using the **Block Editor** ribbon, select the **Close Block Editor** command.

The changes are not reflected in the existing blocks in our drawing file.

6.   Using the **Home** ribbon and the **Block** panel *drop-down list*, and select the **Synchronize Attributes** command. 🖾

7.   Using the Command Line, key in S to **Select** the block in the drawing view.

8.   Use the *Enter* key to answer **YES** to synchronize the ANNO_SECTION_2 block.

The text style change should now be visible in the existing blocks.

In the next section, we will learn how to globally control the drawing objects.

Globally Control Attribute Scales

When working with block Attributes and Annotation scales, blocks can easily be assigned to multiple annotation scales. Removing the unwanted Annotation Scales can be tedious when you finalize your drawing scales. However, did you know you can globally remove the Annotation Scales from these blocks? To accomplish this, just follow these steps:

1.   Continue using the 6-2_Working with Attributes.dwg file.

2.   Using the In-Canvas View Controls, restore the **Custom Model Views | 4-Global Annotation Scale** named view.

3.   Select the three blocks in the drawing view and notice they all have different Annotation Scales and locations defined.

4.   Use the *Esc* key to clear the current selection set.

5.   Select the BOTTOM block and note that it has two Annotation Scales defined.

6.   Use *right-click* to access the **Annotation Object Scale | Add/Delete Scales** command. Using the **Annotation Object Scale** dialog, we can remove any scales that are no longer needed. Select the **1:2** scale and click on the **Delete** button to remove the **1:2** scale.

7.   Click **OK** to save and close the dialog.

8.   This time, select both the UPPER blocks and *right-click* to access the **Annotation Object Scale | Add/Delete Scales** command. It is NOT available for multiple selected blocks.

9.   Instead, keep both blocks selected, and using the **Properties** dialog, locate the **Misc | Annotative Scale** setting and select the **1:1** scale.

10. Using the **Properties** dialog, select the **Add/Delete Scales** icon. 🖾

11. Using the **Annotation Object Scale** dialog, we can remove any scales that are no longer needed. Select the **1:2** and **1:4** scales and click on the **Delete** button to remove both scales.

12. Click **OK** to save and close the dialog.

In the next exercise, we will learn how to change the attribute locations in a single step.

Globally Control Attribute Scale Locations

You will notice the same issue when trying to synchronize the scale locations of multiple blocks. The command does not appear when more than one block is selected. In this case, you can use a ribbon option or key in the synchronization command for more than one block at a time.

ANNORESET	Command Locations
Ribbon	Annotate	Annotation Scaling	Sync Scale Positions
Command Line	ANNORESET (AN)

In this example, we will reset the attribute locations using the SYNC SCALE POSITIONS command:

1. Continue using the `6-2_Working with Attributes.dwg` file.
2. Using the In-Canvas View Controls, restore the **Custom Model Views | 5-Global Location Sync** named view.
3. Select the existing blocks in the drawing view and note the different locations of the various annotation scale blocks.
4. Using the **Annotate** ribbon and the **Annotation Scaling** panel, select the **Sync Scale Positions** command.

In the next section, we will look at some hidden block commands that can assist with using attributes.

Bonus Block Commands

In this section, we will examine how to use the ATTREQ, ATTDISP, ATTIN, ATTOUT, ATTIPEDIT, and ATTWIPE commands to give you more control over your block attributes. These are some of the lesser-known attribute editing commands.

Using the ATTREQ Command

Have you ever needed to place blocks that contain attributes and not had time to input the attribute values right now? If so, use the ATTREQ system variable to disable the attribute value entry to later and delay that task until later.

ATTREQ	
Controls whether INSERT uses default attribute settings during the insertion of blocks. Type: Integer Saved In: Registry	
0	Assumes the defaults for the values of all attributes
1 (default)	Turns on prompts or a dialog box for attribute values, as specified by ATTDIA

Using the ATTDISP Command

Do you need to see all the "invisible" attributes defined in your drawing and display them for review? Use the ATTDISP command to display all attributes, regardless of the flag assigned to the attribute object.

ATTDISP	Command Locations
Ribbon	Home	Blocks	Retain Attribute Display Insert	Block	Retain Attribute Display
Command Line	ATTDISP (ATTD) ATTDISP OFF ATTDISP ON ATTDISP Normal

In this example, we will turn on the display of all block attributes temporarily to review their content:

1. Open the 6-3_Bonus BLOCK Commands.dwg file.

2. Using the In-Canvas View Controls, restore the **Custom Model Views | 1-ATTDISP** named view.

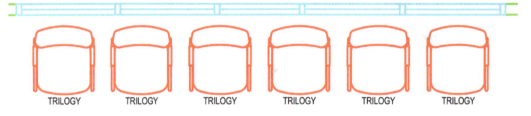

Figure 6.22: Initial Attribute display (retain/normal)

3. Using the **Home** ribbon and the **Blocks** panel, select the *drop-down list* to access the **Display All Attributes** command.

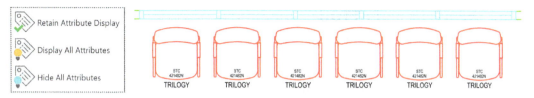

Figure 6.23: Display All Attributes

4. Using the **Home** ribbon and the **Blocks** panel, select the *drop-down list* to access the **Retain Attribute Display** command.

5. Using the command options *drop-down list*, select **Hide All Attributes**.

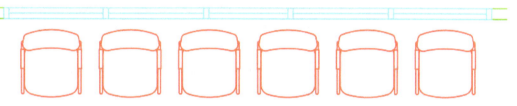

Figure 6.24: Hide Attribute Display

> **Note**
>
> Use this technique to disable the display of attributes that are not up-to-date. Turn OFF all attributes and update them later when you have more time.

These commands use the following system variable to also control the visibility and display of attributes in your drawing.

ATTMODE	
Controls display of attributes.	
Type: Integer	
Saved In: Drawing	
0	OFF: Forces all attributes to be invisible
1 (default)	Normal: Maintains each attribute's current visibility status. Visible attributes are displayed, and invisible attributes are turned off.
2	ON: Forces all attributes to be visible

Using the ATTIN and ATTOUT Commands

The editing process can be very tedious when working with attributes. Did you know that you can export attributes to an Excel spreadsheet, where this data is easier to edit? Here's how:

1. Continue using the 6-3_Bonus BLOCK Commands.dwg file.

2. Using the In-Canvas View Controls, restore the **Custom Model Views | 2- ATTIN_ATTOUT** named view.

3. Select one of the ROOM NAME block objects and *right-click* to select the **Select Similar** command.

4. Using the **Express Tools** ribbon and the **Blocks** panel, select the **Export Attributes** command.

5. Save the attribute data as a new Text file to

   ...\ACAD_TipsTechniques\Exercise_Files\Chapter 06\BLOCK attributes.txt.

6. Start the **Excel** application and open the newly saved TEXT file from the previous step.

7. You can skip the rest of the **Text Import Wizard** and click the **Finish** button.

	A	B	C	D
1	HANDLE	BLOCKNAME	RNAME	RNUM
2	'1E8CD	RoomName	Lobby	124
3	'1E8D9	RoomName	Reception	123
4	'1E8DC	RoomName	Servers	122
5	'1E8DF	RoomName	Break Room	121
6	'1E8E2	RoomName	Supplies	120
7	'1E8E5	RoomName	Supv.	119
8	'1E8EB	RoomName	Mgr.	118
9	'1E8EE	RoomName	Printer	117
10	'1E8F1	RoomName	Conference	116
11	'1E8F4	RoomName	Supv.	115

Figure 6.25: Import Attributes into Excel

8. Using Excel, we can sort the **Room Number** column using the built-in **Sort** command.

9. Select all the data rows and columns, but DO NOT select the HEADER row.

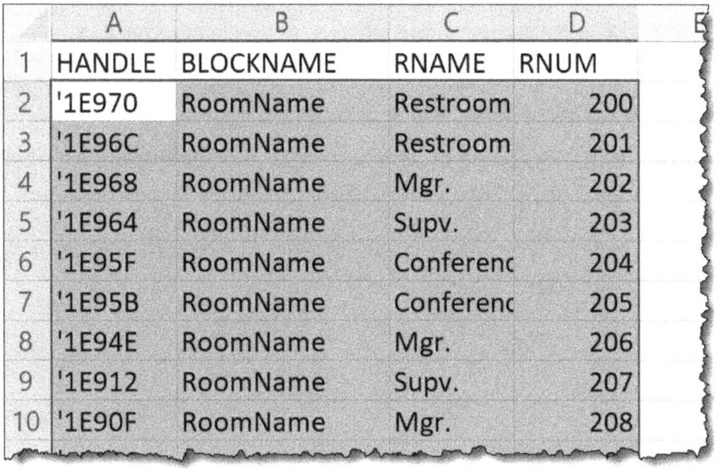

	A	B	C	D	E
1	HANDLE	BLOCKNAME	RNAME	RNUM	
2	'1E970	RoomName	Restroom	200	
3	'1E96C	RoomName	Restroom	201	
4	'1E968	RoomName	Mgr.	202	
5	'1E964	RoomName	Supv.	203	
6	'1E95F	RoomName	Conferenc	204	
7	'1E95B	RoomName	Conferenc	205	
8	'1E94E	RoomName	Mgr.	206	
9	'1E912	RoomName	Supv.	207	
10	'1E90F	RoomName	Mgr.	208	

Figure 6.26: Selecting data in Excel

10. In Excel, using the **Data** ribbon and the **Sort and Filter** panel, select the **Sort** command. Sort

11. Using the **Sort** dialog, set **Sort By** to RNUM and click **OK** to sort the data in numerical order.

12. Change the first-row RNUM value from 100 to 200 in preparation to modify all the room numbers to a 200 series.

13. Select the RNUM 200 data cell and drag the cursor DOWN until all RNUM data is selected.

14. Using the Excel **Home** ribbon and the **Editing** panel, use the **Fill** *drop-down list* and select the **Series** command to auto-increment the values.

15. Using the **Series** dialog, verify that the **Step Value** is set to 1, and click **OK** to close the dialog.

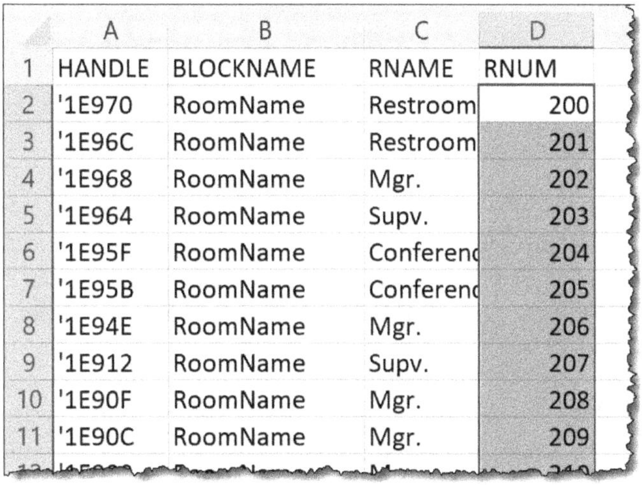

	A	B	C	D
1	HANDLE	BLOCKNAME	RNAME	RNUM
2	'1E970	RoomName	Restroom	200
3	'1E96C	RoomName	Restroom	201
4	'1E968	RoomName	Mgr.	202
5	'1E964	RoomName	Supv.	203
6	'1E95F	RoomName	Conferenc	204
7	'1E95B	RoomName	Conferenc	205
8	'1E94E	RoomName	Mgr.	206
9	'1E912	RoomName	Supv.	207
10	'1E90F	RoomName	Mgr.	208
11	'1E90C	RoomName	Mgr.	209

Figure 6.27: Renumbering data to 200 series

16. Using Excel, select the **Save As** command and overwrite the existing TEXT file created in *Step 5* earlier.

17. Close Excel and return to the AutoCAD application.

18. Using the **Express Tools** ribbon and the **Blocks** panel, select the **Import Attributes** command, select the `6-3_Bonus BLOCK Commands.txt` file, and click the **Open** button.

The room numbers are updated with the new second-floor values. Much quicker than manually editing those room numbers, right?

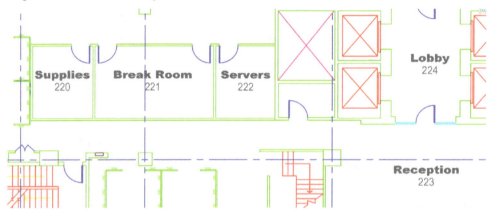

Figure 6.28: Updated room number data results

The next command, ATTIPEDIT, can be used to edit the block attribute without a dialog.

Using the ATTIPEDIT Command

Use the ATTIPEDIT command to edit a block attribute quickly without using the Enhanced Attribute Editor dialog. This allows you to edit the attribute using a simple Text Editor.

ATTIPEDIT	Command Locations
Command Line	ATTIPEDIT (ATTI)

When editing a single-line attribute, the Text Editor is not displayed but if you select a multi-line attribute, the Text Editor is displayed with minimal toolbars and a ruler. You can use a *right-click* to access the TEXT EDITOR | EDITOR SETTINGS to change this default behavior. Let's look at how to use this command:

1. Continue using the `6-3_Bonus BLOCK Commands.dwg` file.

2. Using the In-Canvas View Controls, restore the **Custom Model Views | 3-ATTIPEDIT** named view.

3. Using the Command Line, key in the **ATTIPEDIT** command.

4. *Left-click* on the # Revision Number attribute.

Figure 6.29: ATTIPEDIT single-line attribute

5. Key in 1 for the first revision number.

6. *Left-click* anywhere in the view window to complete the command.

7. Use the *SpaceBar* on the keyboard to recall the **ATTIPEDIT** command.

8. *Left-click* on the **Multi-line Description** attribute.

9. Key in REVISION DESCRIPTION REQUIRING MORE THAN ONE LINE OF TEXT for the revision description.

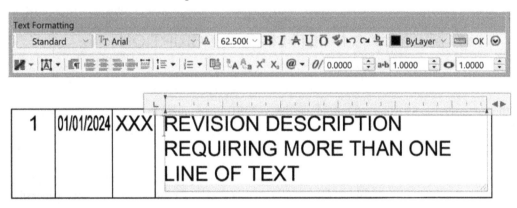

Figure 6.30: ATTIPEDIT multi-line attribute

10. *Left-click* anywhere in the view window to complete the command.

Use the ATTIPE system variable to control the display of the Text Formatting toolbar.

ATTIPE	
Controls which Text Formatting toolbar is displayed when editing multi-line attributes	
Type: Integer	
Saved In: Registry	
0 (default)	Displays the abbreviated Text Formatting toolbar with the Text Editor
1	Displays the full Text Formatting toolbar with Text Editor

> **Note**
> Not all formatting options are available for multi-line attributes, even with the full Text Editor.

Hopefully, everyone knows you can *double-click* on an attribute to edit its value, similar to editing TEXT objects. However, it opens the ENHANCED ATTRIBUTE EDITOR dialog, which is a tedious workflow. Did you know you can automatically run the ATTIPEDIT command if you use the *Ctrl* key when you *double-click* on the attribute? Here's how:

1. Continue using the `6-3_Bonus BLOCK Commands.dwg` file.

2. Using the In-Canvas View Controls, restore the **Custom Model Views | 3-ATTIPEDIT** named view.

3. Use the *Ctrl* key on the keyboard and *double-click* on the **XXX** initials attribute.

4. Key in ABC for the initials and *left-click* anywhere in the view window to complete the command.

Figure 6.31: Automatic ATTIPEDIT command

In the next exercise, we will learn how to use the ATTWIPE command to clear an attribute value.

Using the ATTWIPE Command

If you ever need to clear an attribute value, the previous *Ctrl* shortcut will not remove it, as it requires that some value exist in the attribute. However, there is a free LISP routine out there that will easily CLEAR the attribute value.

Thank you to Chuck Chauvin for this FREE LISP command! I have used this command for years, but it does not work with multi-line attributes. Bummer, right? You can download this FREE LISP command at `https://autocadtips1.com/?s=ATTWIPE` and use it by following these steps:

1. Continue using the `6-3_Bonus BLOCK Commands.dwg` file.

2. Using the In-Canvas View Controls, restore the **Custom Model Views | 4-ATTWIPE** named view.

3. To load a LISP routine in AutoCAD, navigate to the `...\ACAD_TipsTechniques\Exercise_Files\CompanyXYZ\support` folder and *drag-and-drop* the `ATTWIPE.LSP` file into the AutoCAD view window.

4. Using the Command Line, key in `ATTWIPE` and press *Enter* to start the command.

5. *Left-click* on the **Date** attribute to clear the attribute value, and use the *Enter* key to complete the command.

> **Note**
>
> Add any trusted LISP routines to your **Profiles | Trusted Locations** setting to avoid the **Security – Unsigned Executable File** notification.

Summary

In this chapter, we learned how to improve our use of blocks with a clearer understanding of how to use BYLAYER and BYBLOCK. We learned how to better define our block basepoints along with using multiple basepoints to improve our use of blocks. We investigated how to make unexplodeable blocks and how to use some of the older commands that can simplify our daily work with blocks.

In the next chapter, we will look at how to improve our use of blocks by learning how to store our blocks using Block Libraries and how to take advantage of the new Smart Block features.

7

Discover the New Block Tools

In this chapter, you will learn how to use blocks to their fullest by adding the ability to automate daily tasks when using blocks. You will learn how to control blocks when inserting them into your drawings with or without attributes and how to create easy-to-use block libraries and some lesser-known block commands that many of you will find handy.

In this chapter, we'll cover the following topics:

- Using block libraries
- Working with the new smart blocks
- Using the new BCONVERT command

By the end of this chapter, you will be able to use and share your BLOCK objects more efficiently and take advantage of the new smart blocks technology.

Using Block Libraries

What is a Block Library? A Block Library is a single file or folder that contains a collection of standard blocks for your organization. Most users will want to organize these blocks into a single file or folder structure using specific categories to make them easier to find and use. Using a single file is a very efficient way to store similar blocks for easy retrieval by users in the organization.

The single block library file contains individual blocks based on block type. Block files are not different from other drawing files except how they are used. The blocks in this file can contain a graphic image of the block, a block description, the date the block was created, and when the block was last modified. This file can also contain specific instructions for using blocks when needed.

Creating block libraries

First, let's look at how to create a new Block Library:

1. Open the 7-1_Furniture Block Library.dwg file.
2. Using the In-Canvas View Controls, restore the **Custom Model Views | 1-Create Library** named view.

Next, we will populate the Block | Libraries tab.

Populating the Block | Libraries Tab

To populate the Block | Libraries tab, we must decide how to store our blocks. Do you want to keep your blocks as individual files, or do you want to create a library file that contains all of them? You can store them either way, but library files are becoming the de facto standard since we can then keep them more organized, and with the new Legend Table capabilities, they are easier to document.

Next, let's add blocks to the Block Library:

1. Using the **Home** ribbon and the **Block** panel, select the **Insert** *drop-down-list* and select the **Blocks from Libraries** command.
2. Click the **Browse Block Library** button to select either a folder of blocks or a single block library file.

   If you select a folder of block files, the individual block files are read into the Blocks palette, and an image of the block is generated automatically.

   You can add blocks to the Libraries tab using the following methods:

   - *Double-click* on a drawing to view and add all the blocks in the selected drawing.
   - Select the file from the *drop-down-list* to display the most recently used block libraries from either a folder or a single drawing file.
   - Click on **Back to Library** to return to the library and display the blocks and drawings in the folder.

Now that you know how to use the new Block Libraries, when do you use this new feature, and when do you use the previous Tool Palettes feature?

Block Library palette vs. Tool Palette

Why use the BLOCKS | LIBRARIES palette instead of a TOOL PALETTE to insert our blocks? Both have benefits, so let's explore when each is most useful.

Pros

- The BLOCKS | LIBRARIES palette only takes a minute or two to be populated with the blocks in a file
- Any new blocks added to the blocks folder or blocks legend file are automatically added to the BLOCKS | LIBRARIES palette

Cons

- Loading the BLOCKS | LIBRARIES palette is a little slower than the instantaneous TOOL PALETTE
- You can't add commands to the blocks in the BLOCKS | LIBRARIES palette.

> **Note**
>
> The BLOCKS | LIBRARIES palette *drop-down-list* will only remember the last five folders or drawings selected. Eventually, all folders or drawings may drop off this list unless the Block Libraries structure is limited to five folders or drawings.

In the next section, we will examine how to place blocks using the new BLOCKS | LIBRARIES tab.

Placing blocks using the BLOCKS | LIBRARIES Tab

Use either the *click-and-place* or *drag-and-drop* method to insert the block reference:

1. Continue using the 7-1_Furniture Block Library.dwg file.
2. Using the In-Canvas View Controls, restore the **Custom Model Views | 1-Create Library** named view.
3. Using the **Blocks** palette, the options located at the BOTTOM of the palette are used to control the placement, scale, rotation, and automatic repetition of the INSERT command. You can also use the **Explode** option if you want the objects in the block to be inserted as individual objects, not a block.
4. Using the **View** ribbon and the **Palettes** panel, select the **Blocks** palette.
5. Using the **Blocks** palette, select the **Current Drawing** tab and review the blocks in this file.

Once you have populated your Block Library, you must learn how to maintain the libraries using the Block Library dialog.

Removing a Block Library

To remove a block library from the *drop-down list*, you must rename the block library folder. Then, when you select that folder location again in the Block Library *drop-down list*, the folder is found to be missing, and you are presented with a dialog that informs you, **The specified block library was not found. Do you want to remove it from the recent list?** Answer **YES** to remove that folder (library) location from the list.

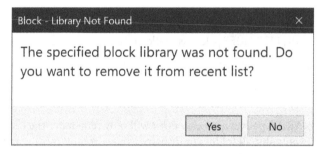

Figure 7.1: Missing Block Library dialog

> **Note**
>
> Deleting the `.blocksMetadata` file from the Block Library folder structure is NOT enough to remove the folder from the Blocks Library drop-down list.

Additional blocks can be placed in a drawing using either the DesignCenter palette or the Tool Palettes, which are covered in later chapters.

Local or Cloud-based

You can also store your blocks in a personal or corporate cloud-based location to gain access to the blocks from anywhere.

> **Note**
>
> You can sync your block libraries to the cloud using the BLOCK INSERT dialog to make your blocks available no matter where you work.

BLOCKSRECENTFOLDER
Defines the path location where recently inserted or created blocks are stored.
Type: String
Saved In: Registry
Initial Value: Varies
To use a cloud storage location, sign in to your Autodesk account to be able to access the recent blocks and block libraries.

BLOCKSYNCFOLDER
Defines the path location where recent and favorite blocks are stored.
Type: String
Saved In: Registry
Initial Value: Varies
To use a cloud storage location, sign in to your Autodesk account to be able to access the recent blocks and block libraries.

BLOCKNAVIGATE	
Defines the folder, file, and blocks that are displayed in the Blocks palette and the Libraries tab.	
Valid values can be folder paths, full filenames of an existing file, or . for none. When set to none, the last-used folder/file used is retained for the next time the palette is opened.	
Type: String	
Saved In: Registry	
Initial Value: Varies	
Example 1	...\ACAD_TipsTechniques\Exercise_Files\CompanyXYZ\block libraries\
Example 2	OneDrive
Example 3	Box, Dropbox, or other Cloud storage devices

In the next section, we will look at how to work with the new Smart Block commands to align and replace blocks.

Working with the new smart blocks

New in AutoCAD 2024 and 2025 is the ability to automatically orient a block during insertion to align with nearby graphics using the new alignment guides. This new smart block functionality can offer placement suggestions based on how you have placed that block in the past.

This type of placement uses AI software to use previous block placements to determine future block placements. Sounds interesting, yes? In reality, I have found that this causes more jumping around and miss-aligned block placements than it saves in time. Your blocks need to be fairly simple for this to work well.

Auto-Align smart blocks

Let me demonstrate a good example, and then I will show you how to turn off the auto-align using the AUTOPLACEMENT and PLACEMENTSWITCH system variables:

1. Open the 7-2_Working with Smart Blocks.dwg file.

2. Using the In-Canvas View Controls, restore the **Custom Model Views | 1-AutoPlacement** named view.

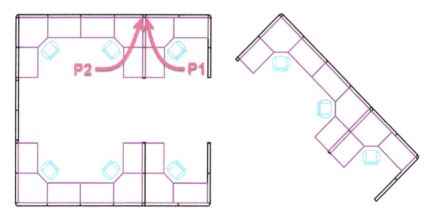

Figure 7.2: Working with Smart Blocks

3. Using the **View** ribbon and the **Palettes** panel, open the **Blocks** palette.

4. Select the **Current Drawing** tab and select the **DESK 2** block.

5. *Hover* over the cubicle wall edge at **P1** to insert the first block as needed.

6. Use the *Esc* key to cancel the current command.

7. Insert the DESK 2 block again and *hover* at **P2** to allow **AutoPlacement** to align the block. A yellow highlight will help you to line up to the correct two cubicle walls. Move the cursor around while "hovering" to achieve the correct block alignment.

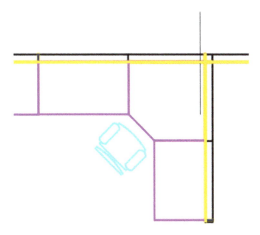

Figure 7.3: Auto-Placement block alignment

8. *Left-click* to accept the block insertion location.

> **Note**
> Use the *Ctrl* key to toggle to other location suggestions or move the cursor away from the current objects to ignore the current suggestions. Use *Shift + W* or *Shift + [* to temporarily turn OFF suggestions while inserting or moving the block.

9. Using the **Blocks** palette and the **Options** settings, turn **ON - Repeat Placement**.

10. Continue placing the **DESK 2** block in the remaining cubicles.

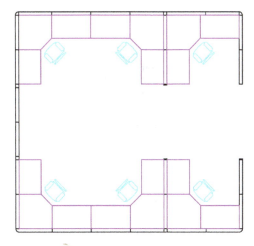

Figure 7.4: Auto-Placement Results 1

11. Practice placing the **DESK 2** block in the angled cubicle walls.

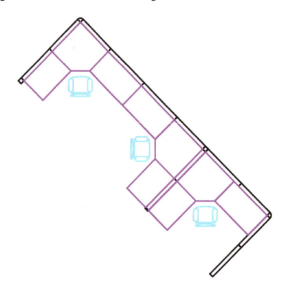

Figure 7.5: Auto-Placement Results 2

Use the following configuration variables to control the Auto-Placement default behavior.

AUTOPLACEMENT	
Controls if the placement suggestions are displayed when you insert a block.	
Type: Integer	
Saved In: Registry	
0	OFF - disables the placement suggestions.
1 (default)	ON - enables the placement suggestions.

PLACEMENTSWITCH	
Indicates if placement suggestions are displayed, by default, when you insert a block.	
Type: Integer	
Saved In: Not Saved	
0	During grip-editing, placement suggestions are OFF by default.
1 (default)	During the INSERT command, placement suggestions are ON by default.

The next exercise will demonstrate using the new REPLACE BLOCK command found in 2024 and higher.

Replace block

Many times, we need to update the drawing file and replace one block with another to reflect design changes in the project. Use the BREPLACE command to easily replace blocks using the new Blocks palette. AutoCAD will provide suggestions for similar blocks when you select a block in the drawing.

BREPLACE	**Command Locations**
Ribbon	Home	Block	Replace
	Insert	Block	Replace
Menu	Misc.	Name	Replace
Command Line	BREPLACE (BREPL)

Using AutoCAD 2023 or later, we can easily replace blocks in the drawing file using the **Properties** dialog:

1. Continue using the 7-2_Working with Smart Blocks.dwg file.

2. Using the In-Canvas View Controls, restore the **Custom Model Views | 2-Replace Blocks** named view.

3. Select one of the round tables with six chairs inside the building and, using the **Properties** dialog, locate the **Misc.** group and, using the **Name** property, select the **Replace** button.

4. Using the **Replace** dialog, select the **Select a Block | Pick** button. Pick

5. Select the round table with 10 chairs located outside the building. The selected table with 6 chairs is replaced with a table with 10 chairs.

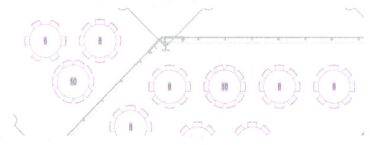

Figure 7.6: Replace Block Individually

6. Select another round table with 8 chairs inside the building, and *right-click* to access the **Select Similar** command. The 8 remaining round tables with 8 chairs are selected (total 8).

7. Again, using the **Properties** dialog, select the **Replace** button.

8. Using the **Replace** dialog, use the **Recent Blocks** list to select the replacement block **TABLE-10 CHAIRS**.

9. Using the **Block – Definition** dialog, select the **Redefine Block** option.

Figure 7.7: Replace Blocks Globally

In the next section, we will look at how to use the new COUNT command to analyze the blocks in your drawing files.

Using the new COUNT COMMAND

There is a new command in AutoCAD 2023 and later: the COUNT command.

COUNT	Command Locations
Ribbon	View | Palettes | Count
Command Line	COUNT (COU)

Count the entire drawing

Let's look at an example of how you can now count the instances of an object or blocks within a specified area. Yes, this will work on more than blocks!

1. Open the `7-3_Using Block COUNT.dwg` file. In this example, there are a few errors in this file that we need to fix. These errors are displayed using the exclamation icons in the COUNT palette. The COUNT command will identify EXPLODED and DUPLICATE blocks in the drawing file.

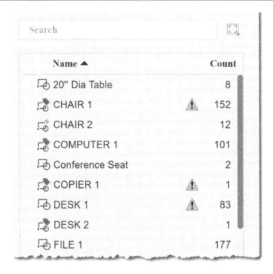

Figure 7.8: COUNT Blocks Errors

2. Using the In-Canvas View Controls, restore the **Custom Model Views | 1-COUNT Blocks** named view.

3. Using the **View** ribbon and the **Palettes** panel, select the **Count** command.

    When the **Count** command is executed, a **Count** palette is automatically opened, and when a block is selected from the **Count** palette, the **Count** toolbar is opened.

    First, let's review the COUNT palette.

4. Using the **Count** palette, review the blocks found in this file and their quantity.

Figure 7.9: COUNT Results 1

Next, let's review the **Count** toolbar.

5.   Using the **Count** palette, select the CHAIR 1 block to open the **Count** toolbar.

Figure 7.10: COUNT toolbar

This toolbar contains the following command features:

- Displays the total number of blocks found in this drawing.   Count: 152

- Left-click on the Details icon to filter the total if needed. ⚠ ⓘ

- Zooms to the NEXT or PREVIOUS object in the count. ⇐ ⇨

- Allows you to define a Count area. ⌞⌟₊

- Select counted objects in Count. 🔼

- Inserts a Count Field object 📊

- Closes the Count toolbar ✕

In this exercise, we will learn to find an exploded block and replace it with a valid BLOCK object.

1.   Select the **DESK 1 Error Icon** ⚠, and a list of the problems appears in the dialog. The LOWER section notifies us that this block has been **Exploded**.

2.   *Left-click* on **Exploded Block** to ZOOM IN to the bad DESK 1. The objects are highlighted in RED, making it easy to identify and fix the problem.

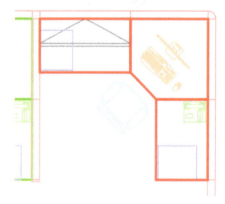

Figure 7.11: Exploded Block Error

3.   Use the *Delete* key or the **Erase** command to delete the exploded graphics for the three RED objects or CUBICLE desks.

4.   Using the **Blocks** palette, select the **DESK 1** block and place the new graphics as a block.

5.   Using the **Count** palette, use the **Back To List** link at the TOP of the palette to return to the Blocks list.

Figure 7.12: COUNT Blocks List

6.   *Double-click* the mouse wheel to execute the **Zoom Extents** command and fit the entire drawing to the current view.

7.   Select the **COPIER 1 Error Icon** and notice it is **Duplicated (Overlapping)**.

8.   Select the **Overlapping object** from the list to ZOOM IN on that object. Again, it is highlighted in RED.

Figure 7.13: Duplicate Blocks

9. Select the **COPIER 1** objects and use the **Properties** dialog to verify that two blocks are indeed overlapping.

10. Use the *Esc* key to cancel the selection set, and use the **Delete** command to remove one of the **COPIER 1** blocks.

11. Using the **Count** palette, use the **Back To List** link at the TOP of the palette to return to the Blocks list.

Use the following system variables to control the COUNT command.

COUNTCHECK	
This system variable is stored as a bitcode using the sum of the following values:	
Type: Bitcode	
Saved In: Registry	
Initial Value: 2	
1	Checks for duplicate objects that overlap on top of each other
2	Checks for duplicate objects that overlap on top of each other, renamed blocks, or exploded blocks

COUNTCOLOR	
Sets the highlighting color on objects in a count.	
Type: Integer	
Saved In: Registry	
Initial Value: 3	
Valid values are from 1 to 255 (AutoCAD Color Index (ACI) colors).	

Be sure to check out the additional system variables in the online help to review more system variables that control the new COUNT command:

- COUNTERRORCOLOR
- COUNTERRORNUM
- COUNTNUMBER
- COUNTPALETTESTATE
- COUNTSERVICE

Count a specified area

In this exercise, we will learn how to count blocks in a specific area:

1. Continue using the 7-3_Using Block COUNT.dwg file.

2. Using the In-Canvas View Controls, restore the **Custom Model Views | 2-COUNT Area** named view.

3. Using the **Count** palette, select the **Count in a Specified Area** button and select all twelve cubicles.

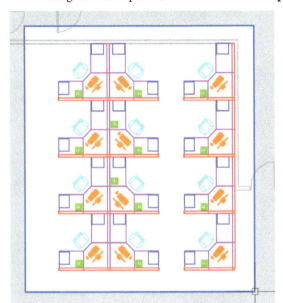

Figure 7.14: Count a Specified Area

4. Using the **Count** palette, select the **Create Table** button located at the BOTTOM of the palette.

5. Now, you have the option to select which blocks you want to include in the new TABLE object.

6. Select the **CHAIR 1**, **DESK 1**, and **FILE 1** furniture blocks only.

7. Using the **Count** palette, select the **INSERT** button.

8. *Left-click* in a blank area of the drawing to place the **BLOCK COUNT** table.

> **Note**
>
> When using the Count command for a specified area, a closed POLYLINE object is placed on the 0-CountArea LAYER for future use. This new layer is locked by default.

The next exercise will demonstrate how to easily access the COUNT command.

Count objects by selection

You can also select objects using normal selection tools and *right-click* to access the COUNT SELECTION command:

1. Continue using the `7-3_Using Block COUNT.dwg` file.

2. Using the In-Canvas View Controls, restore the **Custom Model Views | 3-COUNT Selection** named view.

3. Select the RECTANGULAR POLYLINE DESK object displayed here and *right-click* to access the **Count Selection** command.

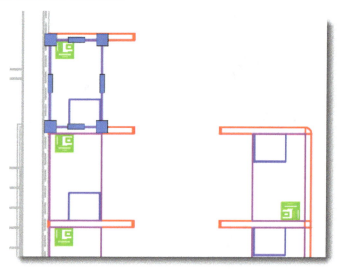

Figure 7.15: Desk Polyline Object

4. The **Count** toolbar opens to display TOTAL matching objects. *Left-click* on the Information button 🛈 to open the **Count** palette if needed.

5. The **Count** palette will display the additional selection options for Match Layer, Match Scale, and Match Mirror State.

Figure 7.16: Initial DESK POLYLINE Selection

6. Using the **Count** palette, turn ON the **Match Scale** option to check that all the POLYLINE DESK objects are on the same scale.

Figure 7.17: Match Scale DESK POLYLINE Selection

Now, you can see that one of the desks has a slightly different scale that is not visually obvious.

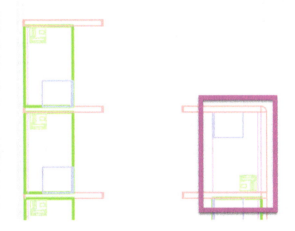

Figure 7.18: Counting Objects Match Scale

7. Delete the miss-scaled DESK object and use the **Copy** command to replace it with another DESK object.

We can also use the COUNT toolbar to navigate to all the matching DESK objects:

1. Using the **Count** toolbar, select the **Next** button ⇨ to navigate to the next matching object. Continue using the **Next** or **Previous** buttons ⇦   ⇨ to visually check each object.
2. Use the **Selected Count** button 🔲 to select the matching 8 DESK objects.
3. Using the **Layer** *drop-down-list*, change the selected objects layer to **A-FURN-WKSF-CALL**.

> **Note**
>
> Only visible objects are displayed in the COUNT palette. The COUNT command does not support all object types in AutoCAD. The following are excluded from the count: text, hatches, 3D objects such as solids and meshes, images, OLE objects, lights and cameras, non-DWG underlays, geographic data, coordination models, point clouds, attribute definitions, external references, construction lines (xlines), and rays.

In the next section, we will learn how to use the new "smart block" BLOCK CONVERT command.

Using the new BCONVERT command

There is a new command in AutoCAD 2025 that allows us to convert similar graphics into blocks automatically. This is one of those features we didn't realize we needed until it was here. What a great feature! I love this kind of surprise!

BCONVERT	Command Locations
Command Line	BCONVERT (BCO)

Convert to Block

Let's look at an example of how you can convert graphics into a block:

1. Continue using the 7-2_Working with Smart Blocks.dwg file.

2. Using the In-Canvas View Controls, restore the **Custom Model Views | 3-Block Convert** named view.

3. Select the RECTANGULAR POLYLINE DESK object displayed here.

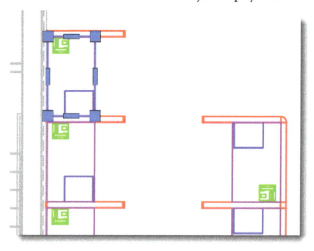

Figure 7.19: Desk Polyline Object

4. Using the Command Line key in the command BCONVERT , and all 9 instances are identifiedin the drawing. You have the choice of converting the following:

   - All nine instances into a single block

   - Source Object only into a single block

   - Individually selected instances into a single block

5. Use the *Enter* key to use the **Select Instances** command option to convert all nine instances to a individual blocks and open the **Convert** dialog.

6. Using the **Convert** dialog, click the **New Block** button.

7. Key in the new block name Call Desk.

8. Click the **Base Point** button  and define the basepoint at **P1**.

Figure 7.20: Convert POLYLINES to BLOCKS

9. Click the **Convert** button to convert all nine instances to blocks and Close the dialog.

Isn't this a great new feature?

Summary

In this chapter, we learned how to improve our use of blocks using the new Block Library capabilities and how to work with the new Smart Block features to auto-align, count, and convert graphics to blocks.

In the next chapter, we will learn how to make dynamic blocks to improve our use of blocks.

8

Learn to Automate with Dynamic Blocks

In this chapter, you will learn how to add intelligent and flexible automation to your old AutoCAD blocks using dynamic blocks. Dynamic blocks can automatically align with other objects, provide multiple basepoints, as demonstrated in *Chapter 6*, control the visibility of block content, and much more, all without exploding the block. You can use dynamic blocks to reduce the number of static blocks in your block libraries.

In this chapter, we'll cover the following topics:

- How to work with dynamic block authoring components
- Improving your work with dynamic examples
- Improving the XYZ COORDINATE block
- Improving the REVISION NOTE block
- Improving the MATCHLINE block
- How to apply chaining to a dynamic block

By the end of this chapter, you will be able to create dynamic blocks to automate common block functionality within your organization.

The parts of a dynamic block

Dynamic blocks contain rules to control the appearance and behavior of a block when it is inserted or modified in a drawing file. These rules can be added to any block to control how the block will be used.

> **Note**
> With dynamic blocks, there are many combinations that produce the same effect. Try to find the simplest solution to achieve the desired results, and do not over-define the dynamic block.

First, let's clarify the main components of a dynamic block and briefly describe how they can be used. A dynamic block can contain parameters, actions, parameter sets, and constraints. It is not necessary to use all of these to improve your block's functionality.

Parameters

Parameters define what geometry is affected by an action in a dynamic block. They visually display grips on the block for users to interact with and can trigger actions to change the block's appearance or contents. A dynamic block must contain at least one parameter.

Actions

Actions are associated with parameters and geometry. They define how the geometry will be modified in a dynamic block when a property or grip changes. Actions do not appear outside the Block Editor and work in the background to manipulate the block contents.

> **Note**
> Use parameters that require no actions, such as point, flip, and visibility parameters.

Parameter sets

Parameter sets are allowed combinations of parameters and actions and can only be used in specific combinations. The following matrix displays the allowed combinations of parameters and actions.

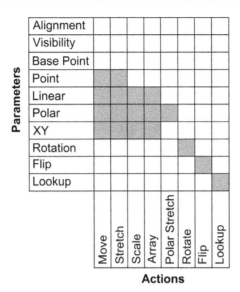

Figure 8.1: Dynamic Block Parameter/Action Combinations

You can also create custom parameter sets with specific property definitions you use most often.

Constraints

Constraints are similar to action parameters but can display custom grips and properties of a block. In place of block actions, you can use constraints to control the geometry in your block.

Block Properties table

The Block Properties table is a newer version of the previous lookup tables. The Block Properties table allows you to define combinations of the custom properties in your dynamic block. You can include the block's properties from parameters, actions, user variables, and block attributes.

I will assume that you know the basics of how dynamic blocks work and are reading this book to learn tips to make you a dynamic block expert. So, let's look at what we can do to improve the use of dynamic blocks.

Using examples to improve your blocks

In this section, we will show several examples of how you can improve static AutoCAD blocks by adding parameters and actions to convert them into more productive dynamic blocks. This section will demonstrate how to improve XYZ Coordinates, Revision Note, and Matchline blocks.

Improving the XYZ Coordinates block

In this exercise, we will learn how to improve the XYZ Coordinate block we created earlier in *Chapter 6*.

First, we want to lengthen the horizontal line between the coordinate values:

1. Open the 8-1_Better Dynamic Blocks.dwg file.

2. Using the In-Canvas View Controls, restore the **Custom Model Views | 1-XYZ Coordinates** named view.

3. Select the **XYZ Coordinate** block and *right-click* to access the **Block Editor** command.

4. Using the **Block Editor** and the **Block Authoring** palette, select the **Parameters** tab.

5. Select the LINEAR parameter and snap to the endpoints of the LEADER and LINE objects in the order shown here as P1 and P2:

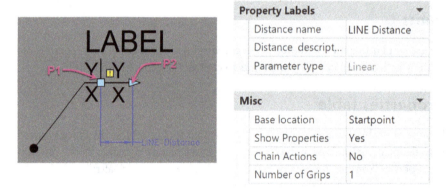

Figure 8.2: XYZ Coordinate LINEAR parameter

6. Select the new LINEAR parameter and using the **Properties** palette and the **Property Labels** section to rename the distance to LINE Distance.

   It is important to name your parameters and actions to "match" their purpose in the block. When dynamic blocks get more complex, you will be glad you got into the habit of naming the contents of the dynamic block.

7. Using the **Properties** palette and the **Misc** section, modify the **Number of Grips** value to 1. We only need one grip since we only want to stretch these objects in one direction.

> **Note**
>
> Don't worry about the warning icon that appears when you have a parameter without an action; this is normal. Not all parameters require an action. We are going to add an action to this LINEAR parameter next.

8. Using the **Block Authoring** palette, select the **Actions** tab and then select the STRETCH action type.

9. Select the **LINE Distance** LINEAR parameter as the parameter to apply this action to, and key in S to define the **Second Point** as the point to associate with this stretch.

10. Draw the **STRETCH Frame** to select both the LEADER and LINE objects, as shown here.

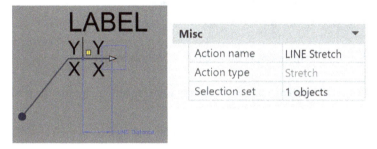

Figure 8.3: XYZ Coordinate LINEAR STRETCH action

11. Select the LEADER object as the object to stretch using the new dynamic definitions and use the *Enter* key to complete the selection.

12. Select the new STRETCH action and using the **Properties** palette and the **Misc** section, modify the action name to LINE Stretch.

13. To review how this action is defined, *hover* over the action icon to display the associated point of the LINEAR parameter, and select the action icon 🔷 to highlight the objects associated with the action.

14. Using the **Block Editor** ribbon, select the **Test Block** command.

15. Select the **XYZ Coordinates** block in the TEST environment and use the exposed grips to test its functionality.

Can you stretch the horizontal line to the preferred length?

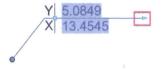

Figure 8.4: Stretch LINE object

16. When satisfied with the dynamic parameters' functioning, select the **Close Test Block** command.

Next, we want to be able to move the coordinate TEXT object if needed:

1. In the **Block Authoring** palette, select the **Parameters** tab and select the POINT parameter type.
2. Snap to the intersection of the lines, as shown here.

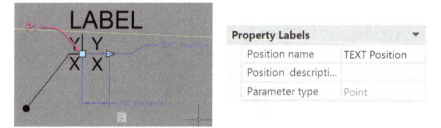

Figure 8.5: XYZ Coordinate POINT parameter

3. Using the **Properties** palette and the **Property Labels** section, modify the position name to
   TEXT Position.
4. In the **Block Authoring** palette, select the **Actions** tab and then select the STRETCH action type.
5. Select the **TEXT Position** POINT parameter as the parameter to apply this action to, and use
   the *Enter* key to accept the second point as the point to associate with this stretch.
6. Draw the **STRETCH Frame** to select all the objects, as shown here.

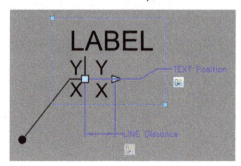

Figure 8.6: XYZ Coordinate POINT STRETCH action

7. Select the new STRETCH action and use the **Properties** palette and the **Misc** section to modify
   the action name to TEXT Stretch.
8. Using the **Block Editor** ribbon, select the **Test Block** command.
9. Select the XYZ Coordinates block in the TEST environment and use the exposed grips to test
   its functionality.

Can you move the XYZ COORDINATE TEXT and do all the lines stretch as expected?

10. When satisfied that the dynamic parameters function as needed, select the **Close Test Block** command and then select the **Close Block Editor** command.

11. Using the **Block – Save Parameter Changes** dialog, select **Save the Changes** to save and close the Block Editor.

Congratulations, you just made your first dynamic block, in this book anyway! The improved functionality demonstrated here with the XYZ Coordinate block can greatly improve a user's ability to label locations in a drawing file.

Next, we want to improve the functionality of the Revision Note block.

Improving the REVISION NOTE block

For this example, we want to add the ability to stretch the linework to accommodate the multi-line TEXT description added in *Chapter 6*:

1. Continue using the `8-1_Better Dynamic Blocks.dwg` file.

2. Using the In-Canvas View Controls, restore the **Custom Model Views | 2-Revision Note** named view.

3. Select the REVISION NOTE block and *right-click* to access the Block Editor.

4. In the **Block Editor**, in the **Block Authoring** palette, select the **Parameters** tab.

5. Select the LINEAR parameter type and snap to the endpoints of LEADER and LINE in the order shown here.

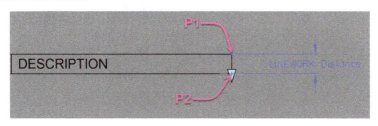

Figure 8.7: Revision Note LINEAR parameter

6. Select the new LINEAR parameter and use the **Properties** palette and the **Property Labels** section to modify the distance name to `LINEWORK Distance`.

7. Continue using the **Properties** palette and the **Value Set** section to modify the following settings:

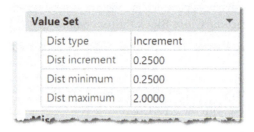

Figure 8.8: Revision Note properties Value Set

The distance type allows you to define an incremental value to grow the linework by the distance of the single-line TEXT attribute (0.2500). The distance maximum allows you to control the longest multi-line TEXT attribute allowed (2.0000). Notice the increment value ghost objects that display after these settings are defined.

8. Continue using the **Properties** palette and the **Misc** section to modify the **Number of Grips** value to 1. We only need one grip to stretch in one direction for this block example.

9. Using the **Block Authoring** palette, select the **Actions** tab and select the STRETCH action type.

10. Select the **LINEWORK Distance** LINEAR parameter as the parameter to apply this action to, and use the *Enter* key to accept the second point as the point to associate with this stretch.

11. Draw the **STRETCH Frame** and use a crossing window to select all the objects except the TOP HORIZONTAL line, as shown here.

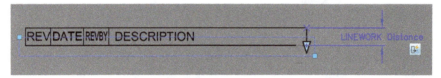

Figure 8.9: Revision Note STRETCH Frame

12. Select the new STRETCH action and use the **Properties** palette and the **Misc** section to modify the action name to LINEWORK Stretch.

13. Using the **Block Editor** ribbon, select the **Test Block** command.

14. Select the REVISION NOTE block in the TEST environment and use the exposed grips to test its functionality.

Can you stretch the linework down to enclose the multi-line ATTRIBUTE text?

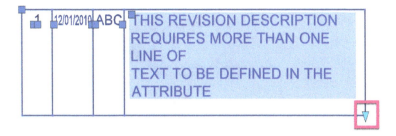

Figure 8.10: Stretch boundary lines

15. When satisfied that the dynamic parameters function as needed, select the **Close Test Block** command and then select the **Close Block Editor** command.

Next, we want to create a more dynamic MATCHLINE block.

Improving the MATCHLINE block

In this exercise, we will create a dynamic MATCHLINE block that will automatically align with our graphics and allow us to stretch the length on both sides simultaneously. This block will use the parameters and actions to lengthen, move, flip, and align graphics in the block.

First, let's look at how this block works; then, we will break it down and see how it was made and make one from scratch:

1. Open the 8-2-Matchline Block.dwg file.

2. Using the In-Canvas View Controls, restore the **Custom Model Views | 1-Matchline NEXT** named view.

3. Select the MATCHLINE_NEXT block and *right-click* to access the **Add Selected** command.

4. *Hover* over the existing lines and notice how the MATCHLINE_NEXT block automatically aligns. You can control the block's alignment by moving the cursor along the line and to each side of the existing line.

5. *Left-click* only when the block is aligned correctly to insert the block.

6. Select the newly inserted block and make note of the available grips.

7. Select the triangle grip ▷ at the end of the MATCHLINE_NEXT block to stretch the length of the block in both directions.

8. Select the arrow grip ⬆ to flip the text to the other side of the MATCHLINE_NEXT block.

9. Select the triangle grip ▷ on the text to separate or move the text away from the sheet number when needed.

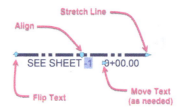

Figure 8.11: Dynamic grips added

Now that you have seen the additional functionality added to this block, let's see how to create one of your own.

Adding a STRETCH parameter to the MATCHLINE_NEXT block

In this exercise we will add a STRETCH action to the LINE object in the MATCHLINE block object.

1. Continue using the 8-2-Matchline Block.dwg file.

2. Using the **Home** ribbon and the **Insert** panel, select the **Insert Block** command and then select the MATCHLINE_NEXT block.

3. Place the MATCHLINE_NEXT block in any blank area of the view window.

4. Select the new block and *right-click* to access the **Block Editor** command.

5. Using the **Block Editor** ribbon and the **Open/Save** panel, use the *drop-down list* to select the **Save Block As** command and name the new block MY_MATCHLINE_NEXT.

The first dynamic condition we want to add is the ability to stretch the length of the line in both directions. This will require a LINEAR parameter and a STRETCH action.

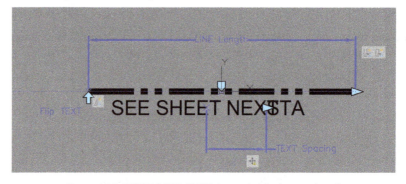

Figure 8.12: MATCHLINE_NEXT dynamic block completed

In this exercise, we will add a LINEAR parameter to allow us to stretch the length of the MATCHLINE_ NEXT linear object.

1. Using the **Block Authoring** palette, select the **Parameters** tab and then the LINEAR parameter, snap to the LEFT endpoint of the LINE object, and then to the RIGHT endpoint of the LINE object.

2. Drag the LINEAR dimension parameter up and *left-click* to place it.

3. Select the new LINEAR parameter and use the **Properties** palette and the **Property Labels** section to change the distance name to LINE Length.

4. Continue using the **Properties** palette and the **Misc** section to modify the **Number of Grips** value to 1.

    Since we are designing this LINE object to stretch in both directions, we only need one grip to stretch the line length.

5. Using the **Block Authoring** palette, select the **Actions** tab and then the STRETCH action type, and then select the **LINE Length** parameter.

6. Select the **LINE Length** "blue" parameter and *left-click* to accept the **RED "X"** at the second point of the LINEAR parameter.

7. Use a crossing window to surround the end of the line, as shown here.

    You only see the crossing window graphics when you *hover* over or select the STRETCH icon.

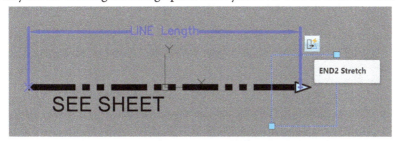

Figure 8.13: LINE Length dynamics for END2

8. When prompted to select objects, use a crossing window identical to the previous step to select the **LINE** object and the **LINE Length** parameter to complete the STRETCH action.

9. Select the new STRETCH icon ⬚ and use the **Properties** palette and the **Misc** section to change the ACTION name to END2 Stretch, for the second point on the LINEAR parameter.

10. Using the **Block Authoring** palette, select the **Actions** tab and then select the **STRETCH** action type.

11. Select the LINE Length "blue" parameter and *left-click* to accept the **RED "X"** at the second point of the LINEAR parameter again to define it as the controlling point for this STRETCH action.

12. Use a crossing window to surround the START end of the line, as shown here.

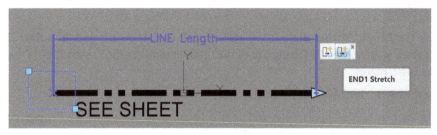

Figure 8.14: LINE Length dynamics for END1

13. When prompted to select objects, use a crossing window to select the **LINE** object and the "blue" **LINE Length** parameter to complete the new STRETCH action.

The new STRETCH icon appears next to the first STRETCH icon placed on the **LINE Length** parameter.

14. Select the new STRETCH icon and use the **Properties** palette and the **Misc** section to modify the action name to END1 Stretch, for the START point on the LINEAR parameter.

15. Continue using the **Properties** palette and the **Overrides** section to modify the **Angle Offset** value to 180.

Now that the LINE STRETCH condition is completely defined, we will test it using the **Test Block** command:

1. Using the **Block Editor** ribbon and the **Open/Save** panel, select the **Test Block** command.

> **Note**
>
> You may need to use the ZOOM EXTENTS command to see the block in the testing environment.

2. Select the MY_MATCHLINE_NEXT block and use the new TRIANGLE grip ▷ to stretch both ends of the line.

3. When satisfied that the dynamic condition functions as needed, select the **Close Test Block** command and then **Close Block Editor**.

Defining attributes in the MATCHLINE_NEXT block

Next, we will add two new block attribute objects to allow the user to key in the exact STATION POINT for the MATCHLINE location and another to capture the value of the current sheet number:

1. Continue using the 8-2-Matchline Block.dwg file.

2. Select the MY_MATCHLINE_NEXT block object and *right-click* to access the Block Editor.

3. Using the **Home** ribbon and the **Block** panel, use the *drop-down list* to access the **Define Attribute** command. 🏷️

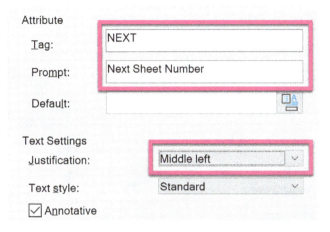

Figure 8.15: MATCHLINE_NEXT Attribute Definition Values

4. Click the **Insert Field** button next to the **Default** field. 🔲

   You can also *right-click* in the field to access the **Insert Field** command.

5. Using the **Field** dialog, change **Field Category** to **SheetSet** and select the **Field Names | CurrentSheetNumber**.

6. Review the **Field Expression** in the lower portion of the dialog and copy (*Ctrl + C*) that expression to the clipboard.

7. Continue using the **Field** dialog to change **Field Category** to **Objects**, and select the **Field Names | Formula**.

8. *Left-click* in the blank **Formula** section of the dialog and paste (*Ctrl + V*) the previous clipboard value.

9. Modify this formula to read `<pasted value>+1`. The `<pasted value>` portion will display as a field value **<pasted formula> +1**.

10. Click **OK** to save these changes and close the dialog.

11. Place the new ATTRIBUTE object so it is aligned with the existing **SEE SHEET** text.

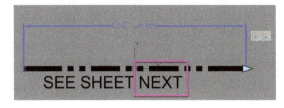

Figure 8.16: NEXT attribute location

Next, we will place the second ATTRIBUTE object for the STATION PT value:

1. Select the **Define Attribute** command again 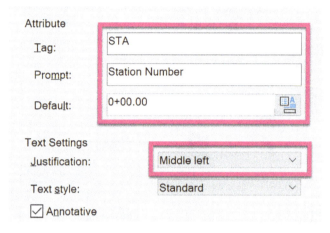 and using the **Attribute Definition** dialog, configure the following values:

Figure 8.17: Tag attribute definition values

2. Click **OK** to save these changes and close the dialog.

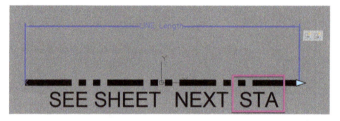

Figure 8.18: STA attribute location

3. Again, place the new ATTRIBUTE object so it is aligned with the existing **SEE SHEET** text as shown here.

Adding MOVE action to the STA attribute

Next, we want to be able to "slide" the STA value back and forth so that it is easily readable after the sheet number attribute is populated; we don't know how many characters that might be. We can do that with another LINEAR and STRETCH parameter set:

1. Continue using the `8-2-Matchline Block.dwg` file.

2. Select the MY_MATCHLINE_NEXT block object and *right-click* to access the Block Editor.

3. Using the **Block Authoring** palette, select the **Parameters** tab and then select the LINEAR parameter. First, snap to the left endpoint of the LINE object, and then snap to the right endpoint of the LINE object.

4. Drag the LINEAR dimension parameter down and *left-click* to place the LINEAR parameter.

5. Select the new LINEAR parameter and use the **Properties** palette and the **Property Labels** section to change the distance name to TEST Spacing.

6. Continue using the **Properties** palette and use the **Misc** section to modify the **Number of Grips** value to 1. Again, we are defining the STA TEXT object to be moveable if needed.

7. Using the **Block Authoring** palette, select the **Actions** tab and then select the MOVE action type.

8. Select the TEXT SPACING parameter and *left-click* to accept the **RED "X"** at the second point of the LINEAR parameter.

9. Select the STA TEXT object as the only object to move when needed.

10. Select the new MOVE icon ⊹ and use the **Properties** palette and the **Misc** section to modify the action name to TEXT Move for the second point on the LINEAR parameter.

11. Using the **Block Editor** ribbon and the **Open/Save** panel, select the **Test Block** command.

12. Select the MY_MATCHLINE_NEXT block and use the new triangle grip to move the STA TEXT object. You should be able to move it in both directions.

13. When you are satisfied that the MOVE dynamic condition functions as needed, select the **Close Test Block** command and then **Close Block Editor**.

The MY_MATCHLINE_NEXT block should look like this when selected.

Figure 8.19: MY_MATCHLINE_NEXT block

Adding a FLIP action to the MY_MATCHLINE_NEXT block

Next, we need to be able to flip the TEXT objects to the other side of the LINE object when needed. We can do this using a FLIP parameter:

1. Continue using the 8-2-Matchline Block.dwg file.

2. Select the MY_MATCHLINE_NEXT block object and *right-click* to access the Block Editor.

3. Using the **Block Authoring** palette, select the **Parameters** tab and then select the FLIP parameter. Then, snap to the left endpoint on the LINE object. This point is the midpoint of the mirror line defined for the FLIP parameter.

4. Drag the cursor to the right at 0 degrees and *left-click* to define the end of the mirror line. When complete, a "blue dotted" mirror line appears.

5. Select the new FLIP parameter and use the **Properties** palette and the **Property Labels** section to change **Flip Name** to TEXT Flip.

6. Using the **Block Authoring** palette, select the **Actions** tab and then select the FLIP action type.

7. Select the FLIP parameter and then select all three TEXT objects when prompted to select objects.

8. Select the new FLIP icon   and use the **Properties** palette and the **Misc** section to change **Action Name** to TEXT Flip.

9. Using the **Block Editor** ribbon and the **Open/Save** panel, select the **Test Block** command.

10. Select the MY_MATCHLINE_NEXT block and use the new arrow grip to flip the TEXT objects to the other side of the LINE object and back.

11. When you are satisfied that the FLIP dynamic condition functions as needed, select the **Close Test Block** command and then **Close Block Editor**.

The MY_MATCHLINE_NEXT block should look like this when selected.

SEE SHEET 1   0+00.00

Figure 8.20: Dynamic FLIP on the MY_MATCHLINE_NEXT block

Adding the ALIGNMENT parameter to the MY_MATCHLINE_NEXT block

Last, we need to easily place the MATCHLINE block in our project and have it automatically align with existing graphics. We can do this using an ALIGNMENT parameter and action:

1. Continue using the 8-2-Matchline Block.dwg file.

2. Select the MY_MATCHLINE_NEXT block object and *right-click* to access the Block Editor.

3. Using the **Block Authoring** palette, select the **Parameters** tab and then the ALIGNMENT parameter.

4. Snap to the 0,0 location and drag the cursor to the right at 0 degrees. Then, *left-click* to define the end of the ALIGNMENT parameter. When complete, a "blue dotted" mirror line appears.

5. Using the **Block Editor** ribbon and the **Open/Save** panel, select the **Test Block** command.

6. Using the **Home** ribbon and the **Draw** panel, select the **Line** command and add two LINE objects at non-orthogonal angles to test the alignment function.

7. Using the **Blocks** panel, select the **Insert Block** command and then select the MY_MATCHLINE_NEXT block.

8. *Hover* over either of the new LINE objects to test whether the block automatically aligns.

9. When you are satisfied that the ALIGNMENT dynamic condition functions as needed, select the **Close Test Block** command and then **Close Block Editor**.

The MY_MATCHLINE_NEXT block should look like this when selected.

Figure 8.21: MY_MATCHLINE_NEXT final results

The MY_MATCHLINE_NEXT block is complete and we can use it to align and automatically number our matchlines in the project.

Using the MY_MATCHLINE_NEXT block

In this section, we will use the new block in both layout and modelspace. A very simple sheet set has been provided to trigger the SheetNumber values:

1. Continue using the 8-2-Matchline Block.dwg file.

2. Using the In-Canvas View Controls, restore the **Custom Model Views | 3-Road Design** named view.

3. Select the **Road Sheet 1** layout and verify that you are in the paperspace environment using the Status Bar.

4. Using the **View** ribbon and the **Palettes** panel, select the **Sheet Set Manager** command. The Sheet Set Manager controls what sheet number is applied to the FIELD in the TEXT object.

5. Insert the MY_MATCHLINE_NEXT block to align with the RIGHT edge of the viewport.

Figure 8.22: MY_MATCHLINE_NEXT in Tab paperspace

6. *Double left-click* inside the viewport to activate modelspace and insert the MY_MATCHLINE_ NEXT block to align with the 20+00 station location.

Figure 8.23: MY_MATCHLINE_NEXT in modelspace

Next, try to make your own MY_MATCHLINE_PREVIOUS block using the In-Canvas View Controls Custom Model Views | 2-Matchline PREVIOUS named view. to automatically populate with the previous sheet number. The completed block (MATCHLINE_PREVIOUS) is provided in the 8-2-Matchline Block.dwg file if needed, but try to create it on your own using the MATCHLINE block again, similar to how we created the MY_MATCHLINE_NEXT block. Enjoy!

The final MY_MATCHLINE_PREVIOUS block will look like this when completed and selected.

Figure 8.24: MY_MATCHLINE_PREVIOUS final results

> **Note**
>
> For this MY_MATCHLINE_PREVIOUS block, you will need to define the previous SheetNumber formula to use the field value of *<pasted formula>-1*.

How to apply the chaining feature to parameters

Chaining parameters refers to using multiple parameters that are controlled by a single action. To accomplish this, we must allow chaining on the parameters. In this example, we will use a series of windows in an elevation view. These windows' height and width will modify using either the "master" window (WINDOW1) or the individual window blocks (WINDOW2 or WINDOW3):

1. Open the 8-1_Better Dynamic Blocks.dwg file.

2. Using the In-Canvas View Controls, restore the **Custom Model Views | 3-Chained Parameters** named view.

First, let's look at how a chained parameter works:

1. Select the **WindowMASTER** block, and using the window on the left (**Window1**), use the horizontal or vertical stretch grips ▷△ to modify the size of all three windows simultaneously.

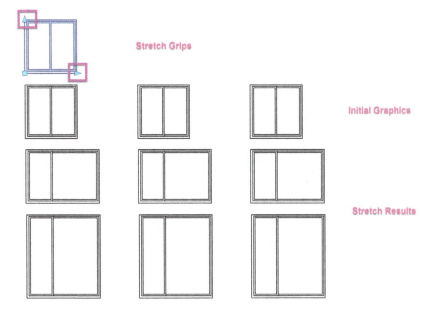

Figure 8.25: Chained dynamic block

Let's look at how chaining a parameter is accomplished:

1. Select the **WindowMASTER** block and *right-click* to access the **Reset Block** command to restore the block to its original size.

2. Using the **Home** ribbon and the **Block** panel, select the **Insert Block** command and insert the **Window1** block.

3. Select the **Window1** block and *right-click* to access the Block Editor.

4. Using the **Block Editor** ribbon and the **Open/Save** panel, select the **Save Block As** command and key in the new block name MY_WindowMASTER.

5. Click **OK** to save the new block and close the dialog.

6. Use the **Copy** command and copy the existing window graphics 60 units to the right and again 120 units to the right.

7. Using the **Block Authoring** palette and the **Parameters Sets** tab, select **Linear Stretch** with one associated linear grip.

8. Place the new LINEAR STRETCH parameter on the bottom edges of the first window.

9. Select the new parameter, **Distance1**, and use the **Properties** palette and the **Property Labels** section to change the distance name to `Window Width`.

10. Select the STRETCH action associated with the WINDOW WIDTH parameter and use the **Properties** palette and the **Misc** section to change the action name to `Width Stretch`.

11. Select the STRETCH action associated with the WINDOW WIDTH parameter and *right-click* to access the **Action Selection Set | New Selection Set** command.

12. Draw a crossing window around the right edge of the window graphics, as shown here, to define the selection set and draw another crossing window in the same location to select objects that will be affected by this grip.

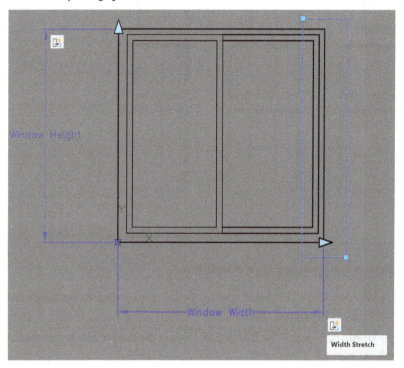

Figure 8.26: Chaining window RIGHT EDGE parameters and actions

13. Using the **Block Authoring** palette and the **Parameters Sets** tab, select the **Linear Stretch** with 1 linear grip associated parameter set.

14. Place the new **Linear Stretch** on the left edge of the first window.

15. Select the new parameter, **Distance2**, and use the **Properties** palette and the **Property Labels** section to change the **Distance Name** to `Window Height`.

16. Select the STRETCH action associated with the WINDOW HEIGHT parameter, and using the **Properties** palette and the **Misc** section, change the action name to `Height Stretch`.

17. Draw a crossing window around the top edge of the window graphics, as shown here, to define the selection set and draw another crossing window in the same location to select objects that will be affected by this grip.

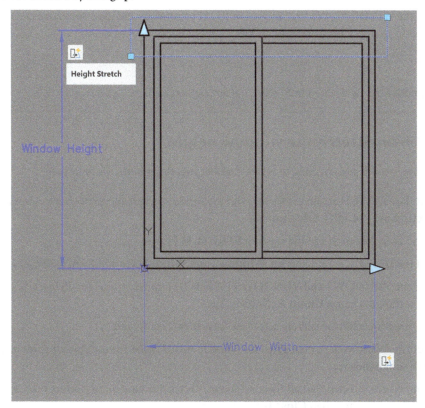

Figure 8.27: Chaining window TOP EDGE parameters and actions

18. Repeat *steps 7 through 17* for the remaining two windows.

Next, we want to add the chaining actions to the last two windows to stretch automatically when the first window is modified.

How to CHAIN Stretch the window width

Adding chained actions to a block allows you to add actions to multiple objects that can be triggered by using a single action grip:

1. Select the STRETCH action associated with the window width and *right-click* to access **Action Selection Set | Modify Selection Set**.

2. Use the *Enter* key to accept the existing **STRETCH Frame**.

3. Select the WIDTH parameters for **Window2** and **Window3** to add them to the selection set. Adding them to the **Window1** selection set links them to any modifications made using this parameter on **Window1**.

4. Select the WINDOW2 and WINDOW3 WIDTH parameters and use **Properties** and the **Misc** section to change **Chain Actions** to **Yes**.

> **Note**
>
> Did you notice the new "chain link" icon that appears on a chained parameter?

How to Chain Stretch the window height

In this exercise, we will add chaining to provide global stretching of the window height:

1. Select the STRETCH action associated with the window height and *right-click* to access **Action Selection Set | Modify Selection Set**.

2. Use the *Enter* key to accept the existing **STRETCH Frame**.

3. Select the HEIGHT parameters for **Window2** and **Window3** to add them to the selection set.

4. Select the WINDOW2 and WINDOW3 HEIGHT parameters and use **Properties** and the **Misc** section to change **Chain Actions** to **Yes**.

5. Using the **Block Editor** ribbon, select the **Test Block** command.

6. Select the **Window1** stretch grips for **Width** and **Height** to test that all three windows stretch simultaneously.

7. When you are satisfied that all three windows stretch simultaneously, select the **Close Test Block** command and then **Close Block Editor**.

> **Note**
>
> Use the RESET BLOCK command to restore the block to its original or default graphics representation.

How to Chain Stretch the middle frame

If you want the middle frames of the window to remain centered, add the following additional stretch actions:

1. Continue using the 8-1_Better Dynamic Blocks.dwg file.

2. Using the In-Canvas View Controls, restore the **Custom Model Views | 4-Middle Frame** named view.

3. Select the block and *right-click* to access the Block Editor.

4. Using the **Block Authoring** palette and the **Actions** tab, select the STRETCH action type.

5. Select the WINDOW WIDTH parameter and draw a crossing window around the MIDDLE WINDOW FRAME of the window graphics, as shown here, to define the selection set and draw another crossing window in the same location to select the window graphics.

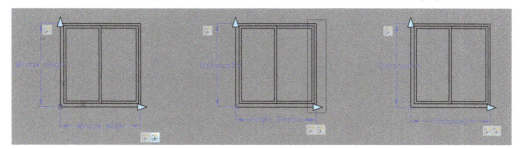

Figure 8.28: Middle frame parameters and actions

6. Select the STRETCH action associated with the WINDOW HEIGHT parameter and use the **Properties** palette and the **Misc** section to change **Action Name** to Middle Stretch.

7. Select the new **Middle Stretch** action and use **Properties** and the **Misc** section to change **Action Name** to Middle Stretch and the **Overrides** section to modify **Distance Multiplier** to 0.5000. A **Distance Multiplier** value of 0.500 applies half of the stretch distance to the middle frame on the windows.

8. Repeat *steps 4 and 7* for **Window2** and **Window3**.

9. Using the **Block Editor** ribbon, select the **Test Block** command.

10. Select the **Window1** stretch grips for **Width** to test that all three windows stretch simultaneously and the middle frames remain centered.

11. When you are satisfied that all three middle window frames stretch simultaneously, select the **Close Test Block** command and then **Close Block Editor**.

This example taught you how to use the "chaining" feature on parameters and how to apply these chains to more than one set of graphics when needed. Be sure to try this "chaining" feature on other parameter types.

Summary

In this chapter, we learned how to improve your blocks and make them more efficient using dynamic tools to apply parameters and actions. Use these same skills and dynamic functionality on other blocks within your organization to improve productivity for everyone!

In the next chapter, we will look at how to improve our use of blocks, including attributes, custom basepoints, block libraries, and groups.

Taking Layers to the Next Level

This chapter will teach you how to manage and use lesser-known layer commands. Managing your layers efficiently is important since we constantly work with the various layer commands throughout our daily workflow.

In this chapter, we'll cover the following topics:

- Tips for using the Layer Properties Manager dialog
- Using overlooked layer commands
- Learning about legacy layer commands

By the end of this chapter, you will be able to manipulate and manage your layers with many tools.

Using the Layer Properties Manager dialog

We are all familiar with using the Layer Properties Manager dialog, but I have found a few features many have never discovered. Let's investigate this dialog further.

Printing the layer List

There are two primary methods for printing a layer list from within AutoCAD. The first method has been available for some time now. You must access the LAYER command from the Command Line, not the Layer Properties Manager dialog.

Method 1: Using the Classic Layer dialog

In this example, we will use the old Classic Layer dialog to print the layer list.

1. Open the 9-1_Using Layer Properties.dwg file.
2. Using the In-Canvas View Controls, restore the **Custom Model Views | 1-Print Layers** named view.

3. Using the Command Line, key in the CLASSICLAYER command to open the previous version of the **Layer Properties Manager** dialog.

4. *Right-click* anywhere in the layer list to access the **Select All** command and use the *Ctrl + C* shortcut to copy the layers to the Windows clipboard. This doesn't appear to do anything, but it does copy the layer information.

5. Open a text file, or better yet, open an Excel spreadsheet and paste the information into the other application.

	A	B	C	D	E	F	G	H	I	J	K	L
1	Name	On	Freeze	Lock	Plot	Color	Linetype	LineWeight	Transparency	Plot Style	New VP Freeze	Description
2	_BorderText	TRUE	TRUE	FALSE	TRUE	30	CONTINUOUS	13	0	Normal	FALSE	Text - Miscellaneous
3	_BorderTitles	TRUE	TRUE	FALSE	TRUE	white	CONTINUOUS	0	0	Normal	FALSE	Plan Text Titles
4	_Clip_Boundary	TRUE	TRUE	FALSE	FALSE	221	DASHDOT2	13	0	Normal	FALSE	
5	_Plot_Shape	TRUE	TRUE	FALSE	FALSE	87	DASHDOT2	13	0	Normal	FALSE	
6	_Plot_Stamp	TRUE	TRUE	FALSE	TRUE	252	CONTINUOUS	0	0	Normal	FALSE	
7	_SheetBorder	TRUE	TRUE	FALSE	TRUE	white	CONTINUOUS	53	0	Normal	FALSE	Sheet Border
8	_SheetLines	TRUE	TRUE	FALSE	TRUE	white	CONTINUOUS	30	0	Normal	FALSE	Sheet Lines
9	_TitleBlock	TRUE	TRUE	FALSE	TRUE	white	CONTINUOUS	0	0	Normal	FALSE	
10	0	TRUE	FALSE	FALSE	TRUE	white	CONTINUOUS	18	0	Normal	FALSE	
11	0-Viewports	TRUE	TRUE	FALSE	FALSE	white	CONTINUOUS	18	0	Normal	FALSE	
12	A-FURNITURE-CHAIR	TRUE	FALSE	FALSE	TRUE	cyan	CONTINUOUS	25	0	Solid	FALSE	
13	BUILDING SHELL	TRUE	FALSE	FALSE	TRUE	151	CONTINUOUS	18	0	Normal	FALSE	
14	CASEWORK	TRUE	FALSE	FALSE	TRUE	cyan	CONTINUOUS	18	0	Normal	FALSE	
15	CHAIRS	TRUE	FALSE	FALSE	TRUE	cyan	CONTINUOUS	18	0	Normal	FALSE	

Figure 9.1: Excel layer list

> **Note**
>
> Once you have the layers in the Excel spreadsheet, you can optimize the columns and rows by selecting all and stretching any "wrapped" columns until they are "unwrapped." Then, *double-click* on a column separator ✛ to optimize the columns and *double-click* on a row separator to optimize the rows.

Method 2: Using the Command Line

You can also get a list of the layers using the non-dialog form of the LAYER command:

1. Continue using the 9-1_Using Layer Properties.dwg file.

2. Using the Command Line, key in the -LAYER command.

3. Select or key in the ? command option and use *Enter* to accept the default * command option to list all the layers in the file. Of course, you can also list all the C* layers if you want just a few of the layers.

Use *Ctrl + C* to select all the layers from the Command Line and paste them into Word or Notepad.

```
1     Layer name        State         Color         Linetype      Lineweight
2  ------------------  -----------   -------------------  ------------  -----------
3  "CASEWORK"            on      -P   4 (cyan)       "CONTINUOUS"  0.180 mm.
4  "CHAIRS"              on      -P   4 (cyan)       "CONTINUOUS"  0.180 mm.
5  "Clipping Boundary"  on           221            "DASHDOT2"    0.130 mm.
6  "COLUMNS"             on      -P   8             "CONTINUOUS"  0.180 mm.
7  "CORE"                on      -P   8             "CONTINUOUS"  0.180 mm.
8  "CPU"                 on      -P   6 (magenta)   "CONTINUOUS"  0.180 mm.
9  "CUBICLE PANELS"      on      -P   6 (magenta)   "CONTINUOUS"  0.180 mm.`
```

Figure 9.2: Layer list text file

> **Note**
>
> You can also use COPY and PASTE in Excel, but you will need to save the layers to a text file and then open the text file in Excel as a delimited file to get the data into various columns.

Method 3: Using XL2CAD – a third-party application

One of my favorite tools for managing layers in AutoCAD is an application from DotSoft (www.dotsoft.com). They offer a tool called XL2CAD that creates a Layer Report in Excel, which includes color icons. This allows you to export and import layers from AutoCAD to Excel so you can use them to fix incorrect layers in a drawing. It's a favorite of mine!

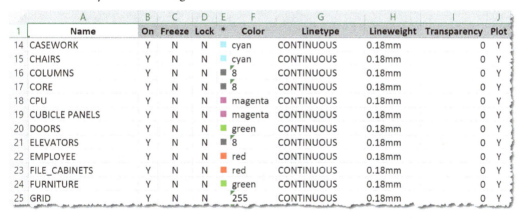

	A	B	C	D	E	F	G	H	I	J
1	Name	On	Freeze	Lock	*	Color	Linetype	Lineweight	Transparency	Plot
14	CASEWORK	Y	N	N		cyan	CONTINUOUS	0.18mm	0	Y
15	CHAIRS	Y	N	N		cyan	CONTINUOUS	0.18mm	0	Y
16	COLUMNS	Y	N	N		8	CONTINUOUS	0.18mm	0	Y
17	CORE	Y	N	N		8	CONTINUOUS	0.18mm	0	Y
18	CPU	Y	N	N		magenta	CONTINUOUS	0.18mm	0	Y
19	CUBICLE PANELS	Y	N	N		magenta	CONTINUOUS	0.18mm	0	Y
20	DOORS	Y	N	N		green	CONTINUOUS	0.18mm	0	Y
21	ELEVATORS	Y	N	N		8	CONTINUOUS	0.18mm	0	Y
22	EMPLOYEE	Y	N	N		red	CONTINUOUS	0.18mm	0	Y
23	FILE_CABINETS	Y	N	N		red	CONTINUOUS	0.18mm	0	Y
24	FURNITURE	Y	N	N		green	CONTINUOUS	0.18mm	0	Y
25	GRID	Y	N	N		255	CONTINUOUS	0.18mm	0	Y

Figure 9.3: XL2CAD report

> **Note**
>
> Please note that while I personally use and recommend this app, I am not responsible for any installation or testing issues that may occur.

Create quick layers

Another quick way to create layers is to use a "," (comma) between layer names in the Layer Properties Manager dialog:

1. Continue using the 9-1_Using Layer Properties 1.dwg file.

2. Using the **Home** ribbon and the **Layers** panel, select the **Layer Properties Manager** command.

3. Using the **Layer Properties Manager** dialog, select the **New Layer** command and key in the layer names Red,Green,Blue.

   When you use the "," comma character, the NEW LAYER command is executed, so you can key in "Red,Green,Blue" to create all three layers.

Using this quick shortcut, you can quickly create new layers when needed.

Displaying longer layer names

This exercise will teach you how to modify the width of the *drop-down-list* Layer to accommodate longer layer names:

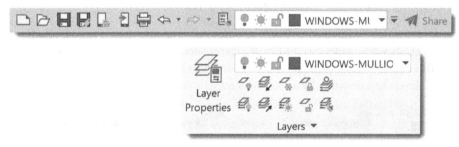

Figure 9.4: Default layer drop-down-list combo boxes

1. Continue using the 9-1_Using Layer Properties.dwg file.

2. Using the **Manage** ribbon and the **Customization** panel, select the **User Interface** command.

3. Using the **Customize User Interface** dialog, select **Workspaces | ACAD_TipsTechniques** and *right-click* to access the **Set Current** command. We do not want to change the out-of-box AutoCAD interface during this course.

   By making **ACAD_TipsTechniques** our current workspace, we can isolate some of the changes made in this course.

   However, changes made to the ribbon apply to all workspaces unless you copy and modify the ribbon parts.

4. Select **Quick Access Toolbars | ACAD_TipsTechniques**, expand it by clicking on the + icon, and select the **Layer** tool.

5. Use the **Properties** section of the dialog in the UPPER RIGHT to modify **Display | Minimum Width** from **180** to **250**.

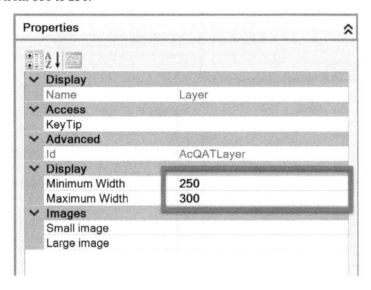

Figure 9.5: Modified QAT layer list combo box

6. Using the **Customize User Interface** dialog, click on the **Apply** button to save these changes and review the size of the **Quick Access Toolbar** (**QAT**) Layer *drop-down-list*. Increase the number defined for **Minimum Width** until your longest layer name displays as needed.

7. When the layer list width is adequate, click the **OK** button to close the dialog.

We can also increase the size of the layer list located in the **Home** ribbon and the **Layers** panel:

1. Continue using the 9-1_Using Layer Properties.dwg file.

2. Using the Command Line, key in CUI to open the **Customize User Interface** dialog.

3. Using the **All Customization Files** section, expand **Ribbon | Panels | Home - Layers | Row 1 | Sub-Panel 1 | Row 1** and select the **Layer Combo Control** tool.

4. Use the **Properties** section of the dialog in the LOWER RIGHT to modify **Display | Minimum Width** from **208** to **250**.

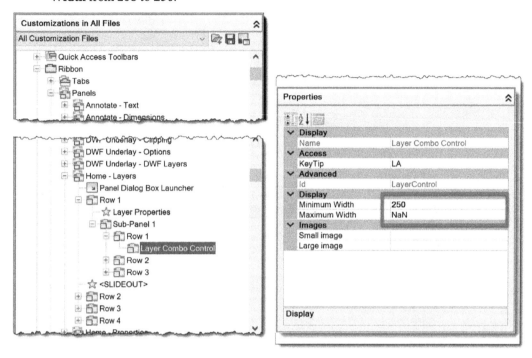

Figure 9.6: Modified ribbon layer list combo box

5. Using the **Customize User Interface** dialog, click on the **Apply** button to save these changes and review the size of the ribbon layer *drop-down-list*.

6. When the layer list width is adequate, click the **OK** button to close the dialog.

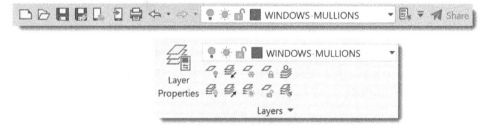

Figure 9.7: Modified layer list drop-down combo boxes

In the next section, we will learn how to organize and modify Layer Properties Manager dialog to define different intent in our drawings.

Other layer tools

Several tools are available to help us manage and use our Layers more efficiently. These include Layer Filters, Layer States, and Viewport Overrides.

Layer filters

In this exercise, we will learn how to define useable Layer Filters to control and minimize our view of the layers in the drawing. Controlling what layers are displayed in the layer list is critical to minimizing your scrolling time when using the layer list:

First, let's define the following Layer Filters:

- Annotation
- Building Structure
- Moveable Items

Next, we will follow these steps:

1. Continue using the 9-1_Using Layer Properties.dwg file.

2. Using the In-Canvas View Controls, restore the **Custom Model Views | 2-Layer Filters** named view.

3. Open the **Layer Properties Manager** dialog and use the **Filters** section of the dialog located on the LEFT side of the dialog.

4. Select **All** layers from the Filter list, *right-click* to access the **New Group Filter** command, and key in the name Annotation.

5. Select **All** layers again, and select the following layers in the list. Use the *Ctrl* key to select multiple layers and *drag-and-drop* them onto the **Annotation** Filter name:

   - EMPLOYEE
   - ROOM NUMBERS

6. Select the **Annotation** filter to verify the layers listed in *step 5* are in the Annotation Group filter.

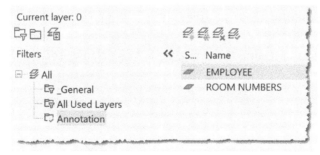

Figure 9.8: Annotation Group filter

Next, we will create another Group filter that contains all the Building Structure layers. To make this easier, we will FREEZE the **Annotation** layers to make it easier to select the other layers:

1.   Continue using the 9-1_Using Layer Properties.dwg file.

2.   Select the **Annotation** filter and select all the annotation layers.

3.   *Left-click* on any of the **Freeze/Thaw** icons ☀ column to FREEZE all these layers.

4.   Select **All** layers from the **Filters** list and *right-click* to access the **New Group Filter** command and key in the name Building Structure.

5.   Select **All** layers from the **Filters** list again and sort the layers based on the FREEZE column. Select all THAWED layers and *drag-and-drop* them onto the **Building Structure** filter name.

6.   Select the **Building Structure** filter to verify all the layers listed in the following image are in the **Building Structure** Group filter.

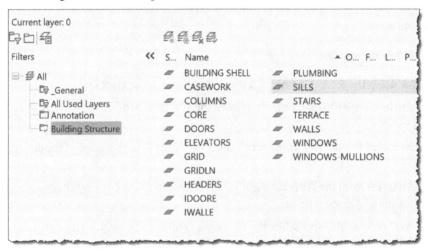

Figure 9.9: Layer Group filter – building structure

Last, we want to create a filter that uses the properties of the layers,

1.   Continue using the 9-1_Using Layer Properties.dwg file.

2.   Select **All** layers from the **Filters** list, *right-click* to access the **New Properties Filter** command, and key in the name Red.

3.   Using the **Layer Filter Properties** dialog, key in **Filter Name:** Red

4. Using the **Filter definition** section of the dialog, *left-click* in the **Color** column and click the **Browse** button ⬚ to select the color **red**.

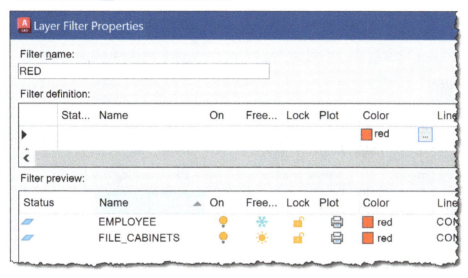

Figure 9.10: Layer Filter Properties – red

5. The LOWER section of the dialog should display two **red** layers, **EMPLOYEE** and **FILE_CABINETS**.

6. Click **OK** to save these changes and close the dialog.

7. Using the **Home** ribbon and the **Layers** panel, select the **Layer Thaw** command 🗐 and thaw all layers in the drawing file.

Now, when working with just **Annotation**, you can select the Annotation layer filter, which will minimize the layer list you must scroll through.

Figure 9.11: Annotation layer filter

> **Note**
>
> Consider using additional Layer Filters to help everyone minimize their scroll time when working with layers.

Now that we have Layer Filters, we can see how to combine them to work with layer States.

Layer States

Use Layer States to take a "snapshot" of the current layer settings or properties such as ON/OFF, FREEZE/THAW, COLORS, and so on. Layer States are saved with the drawing file and can be restored later. You can export and import Layer States between drawing files.

Have you ever needed to create a print or a PowerPoint for a client meeting that highlights specific data in your drawing? Me too! Let's look at how we can change the appearance of our data without permanently changing any of our layer standards.

First, let's look at how to use Layer States:

1. Continue using the 9-1_Using Layer Properties.dwg file.

2. Using the In-Canvas View Controls, restore the **Custom Model Views | 3-Layer States** named view.

3. Open the **Layer Properties Manager** dialog and select the **Layer States** button .

4. Using the **Layer States Manager** dialog, select **Building Structure** and click on **Restore** to apply this Layer State to the current viewport.

Figure 9.12: Building Structure layer filter

5. Using the **Layer States Manager** dialog, select **Building Structure-GRAY** and click on **Restore** to apply this Layer State to the current viewport.

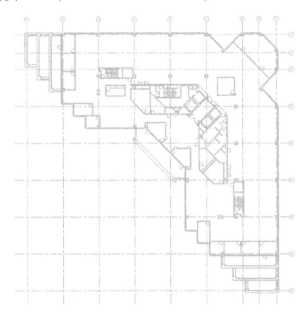

Figure 9.13: Building Structure-GRAY layer filter

Next, let's look at how to create Layer States.

For the first example, we want to grayscale all the permanent building structures and highlight the moveable walls and equipment:

1. Continue using the 9-1_Using Layer Properties.dwg file.

2. Using the In-Canvas View Controls, restore the **Custom Model Views | 3-Layer States** named view.

    Before changing any existing layer standard properties, we need to save the current layer properties in a layer state.

3. Open the **Layer Properties Manager** dialog and select the **Layer States** button .

4. Using the **Layer States Manager** dialog, click **New** and key in the name ORIGINAL to save the current layer properties, and click **Close** to close the dialog.

5. Select the **Annotation** layer filter and **Freeze** all the **Annotation** layers.

6. Using the **Home** ribbon and the **Layers** panel, select the **Layer Freeze** command and select all the "moveable equipment and furniture" in the drawing.

7. Select the **Building Structure** layer filter and select all layers in the list. *Left-click* in the **COLOR** column and change the color to **GRAY** (color 9).

8.   Select **All** layers from the **Filters** list to view all layers in the drawing file.

9.   Sort the **Layer Properties Manager** dialog using the **Freeze** column so that all frozen layers are at the top of the list.

10. Using the **Home** ribbon and the **Layers** panel, select the **Layer Thaw** command [icon] and thaw all layers.

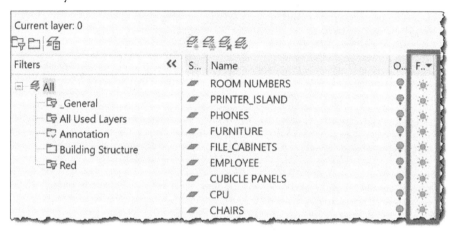

Figure 9.14: Moveable walls and equipment

This provides a clear view of the moveable walls and equipment for client review. We need to save this as another layer state for easy recall later:

1.   Continue using the `9-1_Using Layer Properties.dwg` file.

2.   Using the **Layer Properties Manager** dialog, select the **Layer States** button [icon].

3.   Using the **Layer States Manager** dialog, click **New**, key in the name `Moveable Walls-Equipment` to save the current layer properties, and click **Close** to close the dialog.

Let's use the Layer States we created to recall the Layer Properties as needed:

1.   Continue using the `9-1_Using Layer Properties.dwg` file.

2.   Using the **Layer Properties Manager** dialog, select the **Layer States** button [icon].

3.   Using the **Layer States Manager** dialog, select the **ORIGINAL** layer state and click **Restore** to reset the Layer Properties back to their original definition.

4.   Using the **Layer Properties Manager** dialog, select the **Layer States** button [icon].

5.   Using the **Layer States Manager** dialog, select the **Moveable Wall-Equipment** layer state and click **Restore** to reset the Layer Properties back to the custom layer state.

In this exercise, we learned how to use Layer States to control the display properties of the graphics in our drawing files.

Viewport overrides

You can also apply these Layer States to a specific viewport in a layout:

1. Continue using the 9-1_Using Layer Properties.dwg file.

2. Using the **Layout** tabs located above the Status Bar, select the **Layer States** layout tab.

3. *Double left-click* in the left viewport to activate that viewport.

4. Using the **Layer Properties Manager** dialog, select the **Layer States** button.

5. Using the **Layer States Manager** dialog, select the **ORIGINAL** layer state and click **Restore** to apply this Layer State to the active viewport.

6. *Double Left-click* in the right viewport to activate that viewport.

7. Using the **Layer Properties Manager** dialog, select the **Layer States** button.

8. Using the **Layer States Manager** dialog, select the **Moveable Wall-Equipment** layer state and click **Restore** to apply this Layer State to the active viewport.

   These Layer States are reflected in the Layer Properties Manager using the VP columns in the dialog.

9. Using the **Layer Properties Manager** dialog, review the **VP** columns for the various layer property settings.

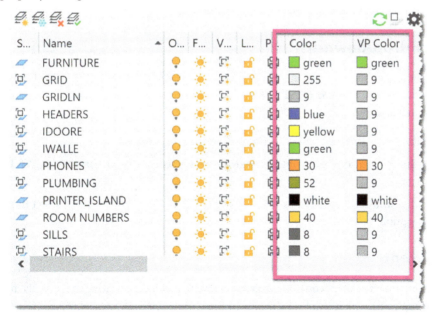

Figure 9.15: Viewport layer properties

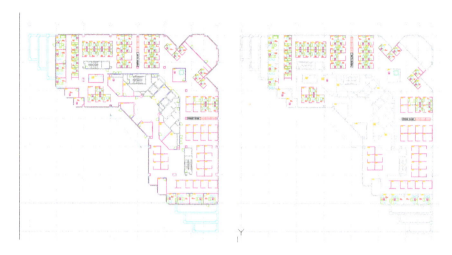

Figure 9.16: Moveable Walls and Equipment Display

Using overlooked layer commands

In this section, we will investigate some of the most overlooked Layer commands, such as in the QAT, or using the Layer ribbon commands such as the MAKE CURRENT, CHANGE TO CURRENT, MATCH, COPY TO LAYER and UNDO LAYER commands.

QAT layer drop-down-list

This simple but very important productivity enhancement will add our CURRENT LAYER *drop-down list* to the **Quick Access Toolbar (QAT)**. This allows you to change your current layer regardless of which ribbon tab is active:

1. Open the `9-2_Overlooked Layer Commands.dwg` file.
2. Using the In-Canvas View Controls, restore the **Custom Model Views | 1-QAT Layers** named view.
3. Using the AutoCAD application title bar and the QAT *drop-down list* ▼ , select the **Layer** display option.

Make Current

Our production drawings often contain hundreds of layers, making navigating the layer list tedious. Scrolling through that long list to find a layer can waste time. Instead, use the MAKE CURRENT layer command to graphically select an object and change the current layer:

1. Continue using the `9-2_Overlooked Layer Commands.dwg` file.

2. Using the In-Canvas View Controls, restore the **Custom Model Views | 2-Make Current** named view.

3. Using the **Home** ribbon and the **Layers** panel, select the **Make Current** command and select a CYAN CHAIR block. This will change the current layer to **CHAIRS**.

LAYMCUR	Command Locations
Ribbon	Home | Layers | Make Current
Command Line	LAYMCUR (LAYM)

Change to Current Layer

You can easily change the layer assigned to an object to another layer using the CHANGE TO CURRENT LAYER command. In this example, a CHAIR1 block is on the wrong layer:

1. Continue using the `9-2_Overlooked Layer Commands.dwg` file.

2. Using the In-Canvas View Controls, restore the **Custom Model Views | 3-Change to Current Layer** named view.

3. Using the **Home** ribbon and the **Layers** panel *drop-down list*, select the **Change To Current Layer** command .

4. Select the **GREEN CHAIR1** block to move the block to the current **CHAIRS** layer. The block will change to the CYAN color since the color is defined as BYLAYER.

LAYCUR	Command Locations
Ribbon	Home | Layers (expanded) | Change to Current Layer
Command Line	LAYCUR (LAYC)

Match Layer

We can use the MATCH LAYER command to move an object from one layer to another:

1. Continue using the `9-2_Overlooked Layer Commands.dwg` file.

2. Using the In-Canvas View Controls, restore the **Custom Model Views | 4-Match Layer** named view.

3. Using the **Home** ribbon and the **Layers** panel *drop-down list*, select the **Match Layer** command .

4. First, select the object you want to change the layer of. In this example, we will select the BLACK/WHITE PHONE block objects.

5. Next, we want to select the object whose layer you want to move the PHONE objects to. We will select one of the ORANGE PHONES.

LAYMCH	Command Locations
Ribbon	Home | Layers | Match Layer
Command Line	LAYMCH (LAYMC)

Copy Objects to New Layer

Next, we will use the COPY TO NEW LAYER command to copy an object from one layer to another. In this example, we want to copy an existing table to the **FURNITURE-RELOCATE** layer.

1. Continue using the 9-2_Overlooked Layer Commands.dwg file.

2. Using the In-Canvas View Controls, restore the **Custom Model Views | 5-Copy to Layer** named view.

3. Using the **Home** ribbon and the **Layers** panel, select the **Copy Objects to New Layer** command.

4. Select the GREEN TABLE block and use the *Enter* key to complete the **Select Objects** prompt.

5. Using the **Copy to Layer** dialog, select the **FURNITURE-RELOCATE** layer.

The table block has been copied to the FURNITURE-RELOCATE layer, and the original block remains on the FURNITURE layer.

COPYTOLAYER	Command Locations
Ribbon	Home | Layers (expanded) | Copy to New Layer
Command Line	COPYTOLAYER (COPYT)

Undo layer changes only

Did you know you can UNDO just your layer changes without affecting other edits? Yes! Using the LAYER PREVIOUS command allows you to undo changes made in the layer list or layer properties without affecting any graphic manipulations.

In this exercise, we will FREEZE a couple of layers and then manipulate the drawing graphics. Using the LAYER PREVIOUS command, we will undo the LAYER FREEZE changes but ignore all graphic manipulations done since the layer changes:

1. Continue using the 9-2_Overlooked Layer Commands.dwg file.

2. Using the In-Canvas View Controls, restore the **Custom Model Views | 6-Layer Previous** named view.

3. Using the **Annotate** ribbon and the **Text** panel, click in the **FIND** field and search the drawing for cubicle number 6022. The view will zoom in to the specific area in the drawing.

4. Using the **Home** ribbon and the **Layers** panel, select the **Layers** *drop-down list* and FREEZE the **FURNITURE** and **CHAIRS** layers.

5. ERASE the CPU and the PHONE equipment in cubicle **6022**.

6. Using the **Home** ribbon and the **Layers** panel *drop-down list*, select the **Layer Previous** command 🐝 to UNDO the CHAIRS layer FREEZE command. Repeat the **Layer Previous** command to UNDO the **FURNITURE** layer FREEZE command. The previous ERASE command performed on the equipment objects is unaffected.

LAYERP	Command Locations
Ribbon	Home | Layers (expanded) | Layer Previous
Command Line	LAYERP (LAYERP)

> **Note**
>
> The LAYER PREVIOUS command will not affect the following layer changes:
>
> - Renamed layers
>
> - Deleted or purged layers
>
> - Added layers

Layer Merge

This exercise will explore various methods for using the LAYER MERGE command:

1. Continue using the `9-2_Overlooked Layer Commands.dwg` file.

2. Using the In-Canvas View Controls, restore the **Custom Model Views | 7-Layer Merge** named view.

In this view, the ORANGE ROOM NUMBERS are on an obsolete layer (RMNUM) and need to be moved to the new layer, ROOM NUMBERS. After they are moved, the obsolete layer (RMNUM) needs to be deleted. If we use the LAYER MERGE command, this can be accomplished in a single step. There are several methods for running this command.

Method 1: Select layers graphically

This method avoids keying in any layer names and allows you to visually select objects:

1. Using the **Home** ribbon and the **Layers** panel, select the **Layer Merge** command 🗂️ and select one of the ORANGE ROOM NUMBER blocks as the **Layer To Merge**.

2. Use the *Enter* key to complete the **Select Objects To Merge** prompt and *left-click* on one of the ORANGE ROOM NUMBERS as the **Target Layer**. The objects are moved and the **RMNUM** layer is deleted.

Method 2: Select layers by name

This method avoids keying in any of the layer names but still provides some visual feedback:

1. Using the **Home** ribbon and the **Layers** panel, select the **Layer Merge** command and select one of the ORANGE ROOM NUMBER blocks as the **Layer to Merge**.

2. Use the *Enter* key to complete the **Select Objects To Merge** prompt and *left-click* on the **Name** prompt to open a list of layers to merge.

3. Select the **ROOM NUMBERS** layer and click **OK** to close the dialog. Then, click **YES** in the **Merge to Layer** dialog to complete the merge.

Method 3: Select layers by dialog

This method avoids selecting an object on the obsolete layer but does not provide graphic feedback:

1. Using the **Home** ribbon and the **Layers** panel, select the **Layer Properties** command and select the **RMNUM** layer.

2. *Right-click* to access the **Merge Selected Layer(s) to…** command.

3. Using the **Merge to Layer** dialog, select the **ROOM NUMBERS** layer and click **OK** to close the dialog. Then click **YES** in the **Merge to Layer** dialog to complete the merge.

Layer Delete

We have all had a layer we couldn't delete because it had something on it, but we couldn't find anything in our drawing that used it! Right? Use the LAYER DELETE command to clean up those drawings where needed. This command will delete all objects using the layer, including blocks that use the layer. So yes, any block definitions that use that layer are also deleted.

But if you have investigated your file and found no objects inside blocks or objects on that layer, this command is a handy tool:

> **Note**
> Before using Layer Delete, consider using the CLEANUP tools to explore your file contents.

In this example, we have a layer that doesn't have anything on it, but cannot be deleted using the PURGE command. So how can we get rid of it?

1. Continue using the `9-2_Overlooked Layer Commands.dwg` file.

2. Using the In-Canvas View Controls, restore the **Custom Model Views | 8-Layer Delete** named view.

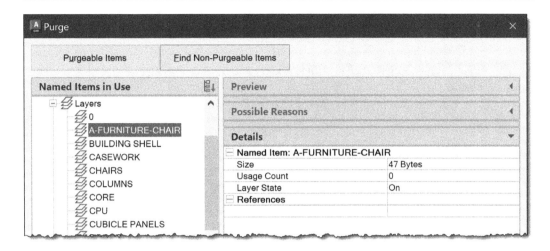

Figure 9.17: Non-purgeable layer

3. Using the **Home** ribbon and the **Layers** panel, select the **Layer Properties** command and select the **A-FURNITURE-CHAIR** layer.

4. *Right-click* on the A-FURNITURE-CHAIR layer name to access the **Delete Layer** command. The command fails since AutoCAD thinks there is an object somewhere in this file that is using this layer.

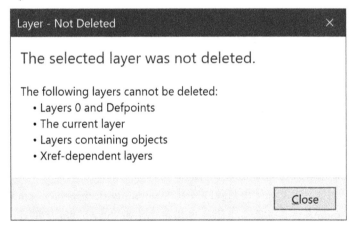

Figure 9.18: Layer delete fails

5. Using the Command Line, key in the LAYDEL command and select the **Name** command option.

6. Select the **A-FURNITURE-CHAIR** layer and click **OK** to close the dialog.

7. Using the **Delete Layers** dialog, click on **YES** to delete the **A-FURNITURE-CHAIR** layer and anything using it.

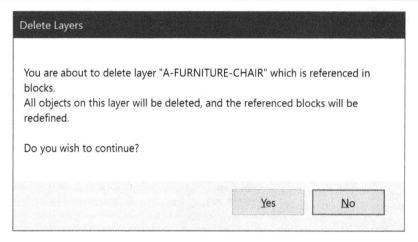

Figure 9.19: Layer Delete dialog

> **Note**
>
> This is also a great way to remove layers and their contents without having to use the LAYER ISOLATE and ERASE commands.

LAYDEL	Command Locations
Ribbon	Home | Layers (expanded) | Layer Delete
Command Line	LAYDEL (LAYD)

Using bonus commands

In this section, we will learn how to use some additional layer commands that might not be easily discovered, such as +LAYER, SETBYLAYER and LAYEREVAL. These bonus commands can be useful for managing your layers.

+LAYER

If you use Layer Filters you will want to know about this little "undocumented" command. Use this command to easily switch layer filters from the Command Line, tool palette, menu, macro, and so on. The +LAYER command allows you to switch to any one of the following layer filter types:

- All Layers
- Properties Filter
- Group Filter
- Unreconciled Layers
- Viewport Overrides

Let's look at how you can use the +LAYER command to more easily utilize Layer Filters:

1.   Open the 9-3_Bonus Layer Commands.dwg file.

2.   Using the In-Canvas View Controls, restore the **Custom Model Views | 1-+Layer** named view.

3.   Using the Command Line, key in +LAYER and review the Layer Filters available in this file.

4.   Key in the layer filter name Annotation to apply this filter.

5.   Using the **Home** ribbon and the **Layers** panel, select the **Layer** *drop-down list* and check that the **ANNOTATION** filter has been applied.

Figure 9.20: Annotation layer filter

SETBYLAYER

Use the SETBYLAYER command to easily fix non-compliant layers inside of blocks without editing them individually:

1.   Continue using the 9-3_Bonus Layer Commands.dwg file.

2.   Using the In-Canvas View Controls, restore the **Custom Model Views | 2-Setbylayer** named view.

     Here, you can see that the CHAIR BYLAYER block inherits the properties of the FURNITURE layer as expected, but the CHAIR2 and CHAIR3 blocks are overridden by properties inside the block definition. I could use the Block Editor to fix this, but the SETBYLAYER command will automatically change all objects inside the block to BYLAYER:

3.   Using the Command Line, key in SETBYLAYER and select all the blocks in this view and use the *Enter* key to complete the selection.

4.   Key in N for **NO** when prompted to **Change BYBLOCK to BYLAYER?**. We wouldn't want to lose our BYBLOCK properties. See *Chapter 6* to learn more about BYBLOCK.

5.   Key in Y for **YES** when prompted to **Include Blocks?**. All blocks are now displayed as expected for the **FURNITURE** layer.

Figure 9.21: SetByLayer initial block graphics

Figure 9.22: SetByLayer block graphics results

> **Note**
>
> Use the SETBYLAYER command on reference files to control your Layer Overrides.

LAYEREVAL

How many times have you heard users complain about the layer notifications? Yep, those notifications quickly become irritating! And what are unreconciled layers?

Unreconciled layers

New layers are added to the drawing file by attaching references and inserting blocks. These new layers trigger a notification balloon, which allows you to control how these new layers display and print. These new layers need to be reviewed if you are using layer overrides or have defined custom plotting settings. Unreconciled layers are automatically added to the Unreconciled Layer Filter for easy identification and review.

You can use the LAYEREVAL and LAYEREVALCTL system variables to control these notifications.

LAYEREVAL	
Determines when the layer list is evaluated for new layers added to the drawing or to attached references.	
Type: Integer	
Saved in: Drawing	
0 (default)	Off
1	Detects when new layers have been added to the attached references.
2	Detects when new layers have been added in the drawing and references.

LAYEREVALCTL	
Controls the Unreconciled Layer Filter in the Layer Properties Manager dialog. Type: Integer Saved in: User settings	
0	Turns OFF layer evaluation and notification of new layers.
1 (default)	Turns ON layer evaluation and notification of new layers for the LAYEREVAL system variable.

LAYERNOTIFY	
Defines when the layer alerts display for unreconciled layers for specific commands. Type: Bitcode Saved in: Drawing	
0 (default)	OFF – no notifications are displayed
1	Plot-notifications display when the PLOT command is executed. `
2	Open-notifications display when the OPEN command is executed.
4	Load/Reload/Attach for references – notifications display when the reference files are loaded, reloaded, or attached.
8	Restore layer state – notifications display when a Layer State is restored.
16	Save – notifications display when the SAVE command is executed.
32	Insert – notifications display when the INSERT command is executed.

> **Note**
> When the LAYEREVALCTL system variable is set to 0, the settings for LAYEREVAL and LAYERNOTIFY are ignored.

In this section, we explored how to use some of the more obscure layer commands such as +LAYER to use layer filters, SETBYLAYER to control blocks and references layer properties, and LAYEREVAL to maintain standard layers in drawing files.

Summary

In this chapter, we learned how to maximize our use of layers by investigating some of the easily overlooked layer features. We learned how to use layer filters, layer states, and layer overrides to take advantage of layers in AutoCAD.

In the next chapter, we will look at how to improve our use of reference files, including PDF and DGN file formats.

10
Enhance Your Knowledge of Reference Files

This chapter will teach you how to use Reference Files (XREFs) to improve data sharing inside and outside your organization. You will learn how to fully understand what capabilities are available and how to improve your reference file performance. You will also learn how to copy multiple objects from a reference and how to use a single reference file more than once. Finally, you will learn how to use PDF data as overlays or as imported data and how to convert the PDF data into useable AutoCAD objects.

In this chapter, we'll cover the following topics:

- Using Attach or Overlay
- Copying objects from a reference or PDF
- Working with duplicate references
- Taking advantage of reference settings
- Improving reference file performance

By the end of this chapter, you will be able to work with DWG and PDF references from within AutoCAD.

Working with reference files

This section will discuss working with various commands to simplify and automate your reference file commands.

Attach or Overlay?

One of the first issues to fully understand is the difference between attaching a reference file and overlaying a reference file.

Attach

When you attach a reference file, this attachment will be visible to anyone who uses your current drawing file as a reference. All "attached" references are automatically inherited for any subsequent attachments; they are permanently visible. Other users will not be able to detach it since you "attached" it as a permanent attachment using the ATTACH method and it is automatically inherited.

Overlay

When you overlay a reference file, the attachment will not be visible to anyone who uses your current drawing file as a reference and it is not inherited automatically. All overlays are forgotten for subsequent reference attachments.

In this exercise, we will learn how to utilize the XREF ATTACH and OVERLAY methods:

1. Open the `10-1_Working with References.dwg` file.

2. Using the In-Canvas View Controls, restore the **Custom Model Views | 1-Attach or Overlay** named view.

3. Using the **Insert** ribbon and the **Reference** panel, select the Dialog Launcher. ↘

4. Using the **External References** dialog, select the **Tree View** button 🗃 to change the view of the attached references so we can visually see how the "attached" references are automatically inherited.

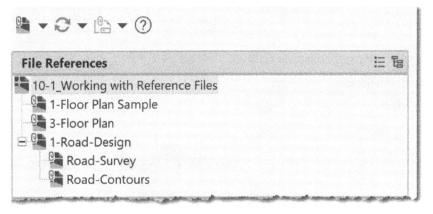

Figure 10.1: References tree view list

5. Open the `Road Design.dwg` file from the `Chapter 10 | Refs` folder.

6. Using the **External References** palette, select the List View button ☰ to change the view of the attached references back to the default list of files.

7. Using the View list, you can see that both the `Road-Survey` and `Road-Contours` files used the **Attach** method, not the **Overlay** method. This causes them to be automatically inherited when the `Road-Design` file is used as a reference.

8. Using the **External Reference** palette and the **Type** column, *double-left-click* on the **Attach** method to change it to the Overlay method. Do this for both the Road-Survey and Road-Contours files.

9. Save and close the Road-Design.dwg file.

10. Return to the 10-1_Working with References.dwg file and, in the pop-up balloon, select the **Reload 1-Road-Design** file link to reload the changes made.

Figure 10.2: Reference Changes Balloon

> **Note**
>
> You can check the **Compare the changes** option in the pop-up balloon to visually review the changes between the previous file and the file being reloaded.

11. Notice that Road-Survey and Road-Design are no longer attached automatically since they are now using the overlay method.

Using XLIST

Have you ever been zoomed in on an area of your drawing and were not able to determine which file, or reference file, an object belongs to? In this case, you could use the XLIST command to display what file an object belongs to:

1. Continue using the 10-1_Working with References.dwg file.

2. Using the In-Canvas View Controls, restore the **Custom Model Views | 2-Using XLIST** named view.

3. Using the Command Line, key in the XLIST command and select a TREE object.

4. The **Xref/Block Nested Object List** dialog opens, showing which reference file or block owns that TREE object.

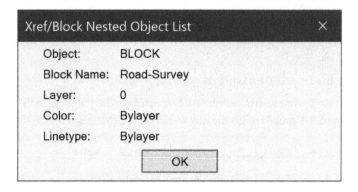

Figure 10.3: XLIST results

Next, let's look at how you can copy objects from a reference file into the active drawing file.

Copying from a Reference

In this exercise, we will examine how to copy objects from the reference file into your active drawing file:

1. Continue using the `10-1_Working with References.dwg` file.

2. Using the In-Canvas View Controls, restore the **Custom Model Views | 3-Copy from Reference** named view.

   One of the first things you'll notice is that the attached reference is "faded." The XDWGFADECTL system variable can be used to control this.

XDWGFADECTL	
Controls the amount of fading or dimming for all DWG reference objects.	
Type: Integer	
Saved in: Registry	
0	The reference file fading is disabled.
> 0	Any positive value controls the fading percentage. Valid values are from 0 to 90.
< 0	Any negative value will disable reference file fading, but the value is saved for easy toggling to that value when the negative sign is changed.
50 (default)	The reference file is faded to 50%.

1. Select the attached reference file, and the **External Reference** contextual ribbon will appear automatically.

2. Using the **External Reference** ribbon and the **Edit** panel, select the **Edit Reference In-Place** command.

> **Note**
>
> You can also *double-left-click* on the reference file to open the **Edit Reference In-Place** command, or *right-click* to access the **Edit Reference In-Place** command.

3. Using the **Reference Edit** dialog, verify the reference file (**Floor Plan Sample**) is highlighted at the TOP of the list and you can choose to open the entire reference file for editing, or just a part of the reference file by selecting objects. Select the **Prompt to select nested objects** option and click **OK** to select the area of the drawing you want to copy objects from.

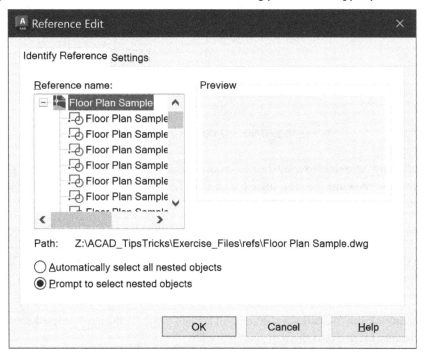

Figure 10.4: Reference Edit dialog

4.   Select the area in the reference file you want to copy objects from. We will select these STAIRCASE objects.

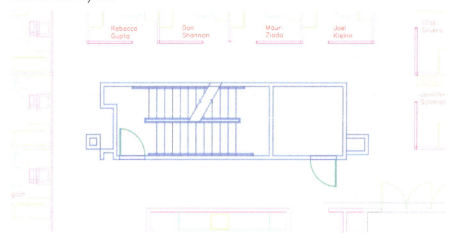

Figure 10.5: STAIRCASE Objects in Reference

5.   Use the **Copy** command to copy the objects to another location or to the same location. I will copy them to the same location since I want the STAIRCASE objects to remain in the same location for all floors in the building.

Next, we need to remove our newly copied objects, not the original objects, from the reference file:

1.   Using the **Active** ribbon and the **Edit Reference** panel, select the **Remove from Working Set** command and key in P for PREVIOUS to select our previous selection set, which is the newly copied objects on TOP of the originals.

     The **Remove from Working Set** command actually removes the newly copied objects from the reference file and puts them in your active file.

2.   Use the *Enter* key to complete the **Select Objects** command option.

3.   Using the **Active** ribbon and the **Edit Reference** panel, select the **Save Changes** command to exit the EDIT REFERENCE command and return to the active drawing file.

4.   Select any of the new objects that are copied into your active file and notice they are essentially a copy from the reference.

Unfortunately, there are no clipping capabilities available during this copy from the reference sequence. Maybe someday, eh?

In the next exercise, we will learn how to attach the same reference file more than once to our active drawing file.

Duplicating References

Many users are not aware that they can attach the same reference file more than once when putting together a detail sheet. In this exercise, we will investigate how to use the process to create a detail sheet.

By default, if you try to attach the same reference file a second time, you will receive the error **<filename> has already been defined** and AutoCAD ignores the attempted reference attachment.

In the next exercise, we will learn how to successfully attach the same reference file more than once.

1. Continue using the `10-1_Working with References.dwg` file.

2. Using the In-Canvas View Controls, restore the **Custom Model Views | 4-Duplicate References** named view.

3. Using the **Insert** ribbon and the **Reference** panel, use the Dialog Launcher ⬛ to open the **External References** dialog.

4. In the **External References** dialog and the **Details** section, rename **Floor Plan** reference attachment Reference Name to **4-Floor Plan Sample**.

5. Continue using the **External References** dialog; *right-click* in the UPPER portion of the dialog and select the **Attach DWG…** command, then select the ...`\ACAD_TipsTechniques\Exercise_Files\Chapter 10\refs\Floor Plan Sample.dwg` file and click the **Open** button.

6. In the **Attach External Reference** dialog, select the following settings:

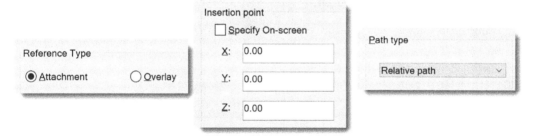

Figure 10.6: Duplicate Reference Settings

7. Click the **OK** button to close the dialog and attach the reference file.

If you need to attach the same DWG file again, just rename the new attachment as needed; in this case, rename it `4-Detail A`.

Note

Renaming a reference allows you to control each reference file independently for layer displays, clipping boundaries, and so on.

You can define a default layer for all references using the XREFLAYER system variable.

XREFLAYER	
Specifies a default layer for all new reference attachments. Type: String Saved in: Drawing	
. (default)	Places the reference attachment on the current layer.
0-XREF	Places the reference attachment on the 0-XREF layer. Use any layer in the current drawing file to hold all references.

> **Note**
> I use the 0-XREF layer as my default reference layer. The "0" keeps the layer at the TOP of the layer list for easy access. Using a specific layer also provides an easy way to turn off the display of all references by FREEZING layer 0-reference file.

Working with Reference Paths

Many times, we need to move drawing files to new locations or new projects. This can cause many of the attached reference file paths to become invalid. Here is how you can fix them:

1. Continue using the `10-1_Working with References.dwg` file.
2. Using the In-Canvas View Controls, restore the **Custom Model Views | 5-Reference Paths** named view.

Removing Reference Paths

In this exercise, we will learn to modify the file path type for attached reference files.

1. Using the **Insert** ribbon and the **References** panel, use the Dialog Launcher ⬋ to open the **External References** dialog.
2. Using the **External References** palette, select the `5-Floor Plan.dwg` reference file.
3. *Right-click* to access the **Change Path Type | Remove Path** command to strip the entire path from the attachment.
4. You can use this same sequence to access **Change Path Type | Make Absolute** or **Change Path Type | Make Relative**.

> **Note**
>
> It is recommended you use the **Options | Files | Project Files Search Path | <project name> | <xref source path>** to define the default location of all reference files for a project. You can define as many path locations as needed.
>
> Use the SET CURRENT button **Set Current** in the OPTIONS dialog to set specific project settings for the current search paths.
>
> Using PROJECTS makes it easier to manage paths when drawings are shared with other users outside our organization who use different drives and paths. When a file is not found at the original path, the project paths associated with our PROJECTNAME are searched. If PROJECTNAME paths are not found, AutoCAD uses its default search paths.

5. Close the `10-1_Working with References.dwg` file and save your changes.

6. Re-open the `10-1_Working with References.dwg` file and notice that the reference file is **Not Found** as expected since we removed the path for this reference attachment and no PROJECTNAME is defined.

   We have all seen this error more often than we would like!

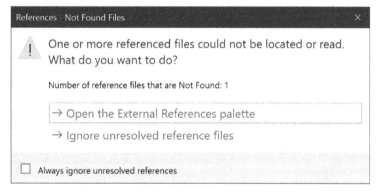

Figure 10.7: Reference Not Found

7. Select **Ignore unresolved reference files**.

   An easy fix when we have our PROJECT PATHS defined is to use the PROJECTNAME command to set project paths.

8. Using the Command Line, key in `PROJECTNAME`. Your current PROJECTNAME setting is probably defined as NONE. Key in the Projectname `ACAD_TipsTechniques`.

   This PROJECTNAME has been defined in the **Profile** delivered with the course and installed in *Chapter 1*. It can be reviewed using **Options | Files | Project Files Search Path | ACAD_TipsTechniques**.

Figure 10.8: ACAD_TipsTechniques project paths

9.   Close the 10-1_Working with References.dwg file and save your changes.

10. Re-open the 10-1_Working with References.dwg file and notice that the reference file is found as needed.

Let's look at how the PROJECT NAME SEARCH PATHS are defined:

1.   Continue using the 10-1_Working with References.dwg file.

2.   *Right-click* anywhere in the drawing view and select the **Options** command.

3.   In the **Options** dialog, select the **Files** tab and expand **Project Files Search Path | ACAD_TipsTechniques**.

Using these PROJECT SEARCH PATHS, your references will now search for references using the following workflows.

Workflow 1: No Path

Using the NO PATH option is useful if your workflow regularly moves drawing files around or you don't have a defined project folder structure.

When the reference attachment has no path defined, AutoCAD searches using this workflow:

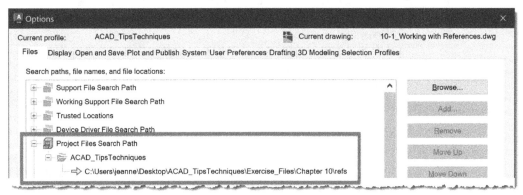

Figure 10.9: Options | Project Files Search Path

No Path Workflow
1. Current drawing folder
2. Project Search Paths (PROJECTNAME)
3. Options Search Paths
4. Start In folder defined in Windows

Workflow 2: Relative Path

Using the RELATIVE PATH option is useful if your workflow regularly moves drawing files around or you don't have a defined project folder structure.

When the reference attachment has a relative path defined, AutoCAD searches using the following syntax rules:

Path Syntax	Description
\	Look in the root folder of the host drawing.
path	Look in the host drawing folder and follow the specified path.
\path	Look in the host drawing folder and follow the specified path.
.\path	Look in the host drawing folder and follow the specified path. (.) always refers to the current folder.
..\path	Look in the folder that is one level higher than the host drawing folder and follow the specified path. (..) refers to a folder that is one level higher than the current folder.
..\..\path	Look in the folder that is two levels higher than the host drawing folder and follow the specified path. (..\..\) refers to a folder that is two levels higher, or the grandparent folder, of the current folder.
Using these path syntax rules allows a file to be moved between drives using the same folder structure.	

Use the REFPATHTYPE system variable to control your default reference attachment path type.

REFPATHTYPE	
Controls whether reference files are attached using full, relative, or no paths when attached to a drawing file. Type: Integer Saved in: Registry	
0	No path is used for the reference attachment
1 (default)	Relative path is used for the reference attachment
2	Full path is used for the reference attachment

Reference Manager

Have you ever received files from someone outside of your company and all the references, fonts, linetypes, or other resources were missing?

Many times, we find several files with invalid reference attachments that need to be repaired or modified to a new path location. This is easily accomplished using the Reference Manager delivered with AutoCAD.

The Reference Manager allows you to modify a list of selected drawings with unresolved resources or reference attachments. Use this utility to fix references, text fonts, linetypes, standards, images, plot configurations, and plot styles:

1. The Reference Manager is a standalone application that can be found using the Windows **Start menu | AutoCAD 2025 | Reference Manager**.

2. In the **Reference Manager** dialog, select **View | Options** and turn ON the **Show full path in tree view** option.

3. Verify that **Profile used to resolve references** is set to **ACAD_TipsTechniques**.

4. Turn ON the **Add xrefs when adding drawings** option.

5.   Click **OK** to save these changes and close the dialog.

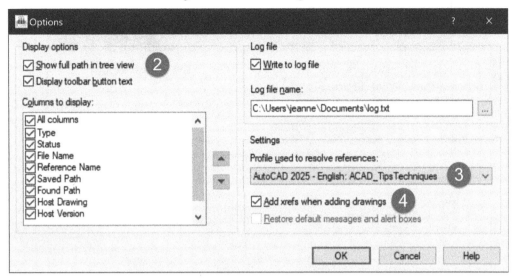

Figure 10.10: Reference Manager Options

6.   In the **Reference Manager** dialog, click the **Add Drawings** button.

Add the following drawing files:

- `10-2_Furniture Plan.dwg`

- `10-2_Lighting Plan.dwg`

- `10-2_Power Plan.dwg`

- `10-2_HVAC Plan.dwg`

7. In the **Reference Manager - Add Xrefs** dialog, select the **Add all xrefs automatically regardless of nesting level** option.

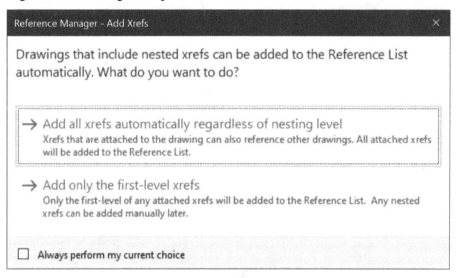

Figure 10.11: Reference Manager - Add Xrefs

8. Using the Reference Manager, select **All Drawings** at the TOP of the list and *right-click* to access the **Expand Branch** command to view all items for all the files listed. *Right-click* again to access the **Collapse Branch** command to minimize the list.

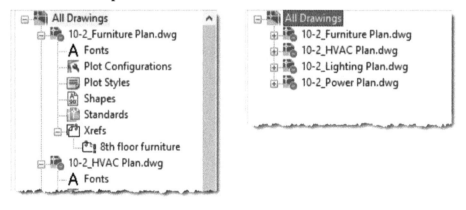

Figure 10.12: Reference Manager File List

9. Continue with **All Drawings** selected. Using the RIGHT side of the dialog, select all of the reference file attachments and *right-click* to access the **Edit Selected Paths...** command.

10. In the **Browse for Folder** dialog, select the Chapter 10 | refs folder. Click **OK** to close the dialog, and click **OK** again to complete the command.

11. Continue with **All Drawings** selected. Using the RIGHT side of the dialog, select all of the reference file attachments and *right-click* to access the **Find and Replace ...** command, which will find specific paths and replace them with updated paths.

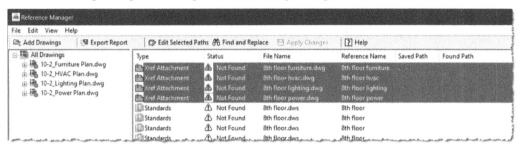

Figure 10.13: Edit Selected Paths

Don't forget to return to and use this utility for all your missing resource problems.

The next exercise will demonstrate how to find where a DWG file is being used as a reference file.

Finding References

How many times have you needed to move or rename a drawing file, and you need to find where it has been referenced? You can use the DesignCenter to search for references based on their filename.

In this example, we want to find out what files are using 9TH FLOOR PLAN.DWG as a reference:

1. Open the 10-1_Working with Reference Files.dwg file.
2. Using the **View** ribbon and the **Palettes** panel, select the **DesignCenter** command.
3. Select the **Search** button and set **Look for** to **Xrefs**.
4. Use the **Browse...** button to navigate to the Exercise_Files folder for this chapter.
5. Search for the 9TH FLOOR PLAN filename and click the **Search Now** button.

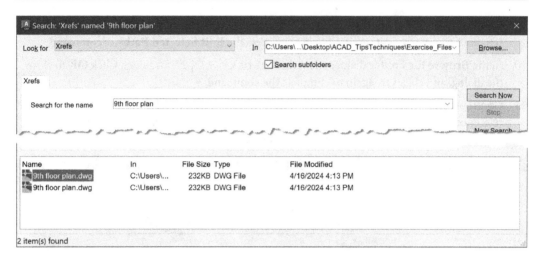

Figure 10.14: Xrefs Search results

The results confirm that there are two files that are referencing the 9TH FLOOR PLAN.DWG file.

Reference Balloon Notifications

There are several system variables to help you control the notification balloons for your personal use of reference files.

XREFNOTIFY	
0	Notification of missing references is disabled
1	Notification of missing references is enabled
2	Notification of missing references and modified references is enabled

You can control the length of time the notification balloon is displayed using the XNOTIFYTIME system variable.

XNOTIFYTIME
Defines the display time for the reference file notification balloon in minutes. This is the number of minutes between checking for modified reference files.
Type: Integer
Saved in: Registry
Using the Command Line, you must key in the following: (setenv "NOTIFYTIME" "n") where "n" is the number of minutes between 1 and 10,080 (seven days). The default value is 1 minute. Use the key in (getenv "XNOTIFYTIME") to view the current setting.

In the next exercise, we will look at how to convert a reference file to a block object.

Converting a reference file to a Block

Have you ever needed to convert a reference file to a block so that all data is contained in a single file for archival or transfer purposes? The following steps make this easy:

1. Continue using the `10-1_Working with References.dwg` file.
2. Using the In-Canvas View Controls, restore the **Custom Model Views | 6-Convert XREF to Block** named view.
3. Using the **Insert** ribbon and the **References** panel, use the Dialog Launcher ⬂ to open the **External References** dialog.
4. Using the **External References** dialog, select the `6-Floor Plan.dwg` reference file and *right-click* to access the **Bind** command.
5. Using the **Bind/DGN Underlay** dialog, select the **Insert** option and click **OK** to close the dialog.
6. Select the new BLOCK object that was created using the selected reference file and review the object using the **Properties** dialog. The name of the block is generated using the name of the reference file.

In the next exercise, we will convert a block to a reference file.

Converting a Block to a reference file

Do you work with users who tend to BIND the references into the current drawing to simplify file sharing with others outside your organization? Yes, this is a common workflow, but it is not the correct way to handle the file exchange. However, it happens, and we need to be able to restore the reference files.

Use the following steps to restore a reference file:

1. Continue using the `10-1_Working with References.dwg` file.
2. Using the In-Canvas View Controls, restore the **Custom Model Views | 7-Convert Block to XREF** named view.
3. Using the **Express Tools** ribbon and the **Blocks** panel, use the **Blocks** *drop-down list* and select the **Convert Block to Xref** command.
4. Use the **Pick** button to select the block to convert.
5. Using the **Select an Xref File** dialog, select the original drawing `New Floor Plan.dwg` file, click **Open** to close the dialog, and convert the block back to a reference file.
6. Use the *Enter* key to accept the default value **YES** to **Purge unreferenced items when finished**.

> **Note**
> I have experienced some issues with this command in AutoCAD 2025, but I have used it with no problems in previous versions.

In the next section, we will discuss how to improve your drawing performance using reference files.

Improving Reference Performance

With our drawing files increasing in size, it can be important to improve the performance of the reference file attachments when needed. You can improve the performance of the reference file using LAYER and SPATIAL INDEXING, especially when clipping a small area in a reference file or freezing several layers of the reference file.

Using Layer and Spatial Indexing

By default, references in AutoCAD are defined to use demand loading with the ENABLED WITH COPY option, as defined in the OPTIONS dialog.

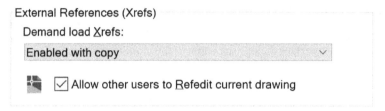

Figure 10.15: References Demand Loading

Indexing can reduce the amount of data loaded when you attach a file as a reference. For example, if you know the file you are creating will be used as a reference, you can save that file with LAYER and/or SPATIAL INDEXING. To get the maximum benefit out of demand-loaded reference files, it is recommended that you save drawings that will be used as reference files with layer and spatial indexes.

Layer Indexing

A layer index is a list defining what objects are on which layers. This list is used when the drawing is referenced using demand loading to determine which objects must be displayed. Objects on frozen layers in a reference file are not read if the reference file has a layer index and is being demand-loaded.

Spatial Indexing

A spatial index organizes drawing objects based on their location in the file. This index determines which objects need to be displayed when the drawing is referenced using demand loading.

Using these indexing options allows the drawing to display only the objects and layers that are needed based on the reference clip boundary.

Drawing files that are not used as references or partially opened do not benefit from these indexing settings.

Let's look at how you can define this indexing in your drawings:

1. Open the `10-3_Furniture Plan.dwg` file.

2. Use the **File | Save As** command to open the **Save Drawing As** dialog and select the **Tools** *drop-down list* to access the **Options** command. Tools ▼

3. Using the **Saveas Options** dialog, modify **Index type** to **Layer & Spatial**.

4. Click **OK** to save this change and close the dialog.

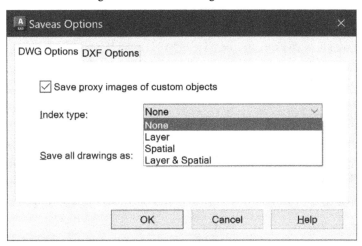

Figure 10.16: Layer and Spatial Setting

5. Click the **Save** button and click **Yes** to replace the existing file.

There is no significant change in the file size when using this indexing.

INDEXCTL	
Defines which indexing is created, either Layer, Spatial, or both, in a drawing file.	
Type: Integer	
Saved in: Drawing	
0 (default)	No indexing is created
1	Layer indexing is created
2	Spatial indexing is created
3	Layer and Spatial indexing is created

In the next section, we will look at using PDF files as references.

Working with PDFs

You can ATTACH or IMPORT a PDF in AutoCAD. The difference is whether you want an image reference file or editable objects. In this section, we will look at how to ATTACH, IMPORT, and CLIPBOARD COPY from a PDF.

PDF Clipboard Copy

Hopefully, all of you know that you can attach a PDF as a reference file and do all the same manipulations you would with a DWG reference. Many times, I would attach the PDF and then CLIP the portion I wanted. But did you know you can use the CLIPBOARD to COPY and PASTE just the portion of the PDF you want?

1. Open the `10-4_Working with PDF Files.dwg` file.

2. Using the In-Canvas View Controls, restore the **Custom Model Views | 1-PDF COPY** named view.

3. Open the `10-4_Truss_Roof_Details.pdf` file using Acrobat Reader.

4. Using the PDF file, *left-click and drag* a selection box around the **A5** detail.

5. **Zoom In** on the **A5** detail and *right-click* to access the **Take a Snapshot** command.

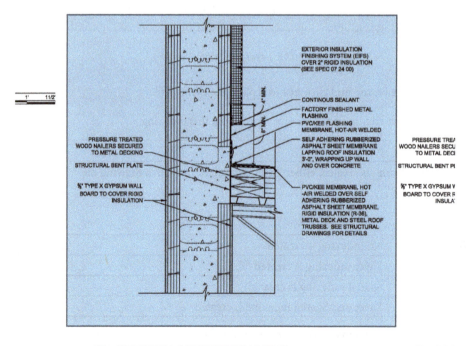

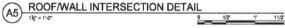

Figure 10.17: PDF Selection Area

> **Note**
> Be careful not to include the edges of the PDF in your selection box, or you will see that edge when you print from AutoCAD. AutoCAD's *FRAME system variable settings cannot control those edges. You need to ZOOM IN as close as possible to get a good image resolution.

6. Return to AutoCAD and use *Ctrl + V* to paste the image into your file.

If you want to control the size of the "image" manually, you must use **Clipboard | Paste Special | Image Entity** to paste the image into the drawing file. I generally paste the image and then use SCALE BY REFERENCE as needed.

To control the plotting of edges in AutoCAD for images, we need to use the IMAGEFRAME system variable. I prefer to use a value of **2** so that the edges of the image are visible and easier to select if I need to move the image in the drawing file.

IMAGEFRAME	
Controls the display and plotting of image frames. Type: Integer Saved in: Drawing	
0	The image frame will not display or plot. The image frame will temporarily appear when the image is selected.
1 (default)	The image frame will display and plot.
2	The image frame will display but does not plot.

PDFFRAME	
Controls the display and plotting of PDF underlay frames. Type: Integer Saved in: Drawing	
0	The PDF frame will not display or plot. The image frame will temporarily appear when the image is selected.
1 (default)	The PDF frame will display and plot.
2	The PDF frame will display but does not plot.

FRAME	
Controls the display of frames for all images, map images, underlays, clipped references, and wipeouts. Type: Integer Saved in: Drawing	
0	The object frame will not display or plot.
1	The object frame will display and plot.
2	The object frame will display but will not plot.
3 (default)	The object frame varies for all objects with frames – images, underlays, clipped references, and wipeouts – and they do not have the same frame settings. Note: You cannot manually set the FRAME system variable to a value of 3. The setting of 3 is only restored when one of the object-specific frame settings is changed to a value that is different from the others.

> **Note**
>
> The FRAME system variable overrides the IMAGEFRAME setting. To reset the image frame settings, set the IMAGEFRAME system variable after the FRAME system variable.

FRAMESELECTION	
Controls the selection of frames for an image, underlay, clipped xref, or wipeout. Type: Integer Saved in: Registry	
0	Hidden frames cannot be selected
1 (default)	Hidden frames can be selected

Of course, another great method for using PDF data is to import the PDF as actual AutoCAD objects. In the next exercise, we will import PDF data from a multi-page PDF to use as editable objects.

Importing a PDF

In this exercise, we will import the PDF rather than attach it as an image or reference:

1. Continue using the `10-4_Working with PDF Files.dwg` file.

2. Using the In-Canvas View Controls, restore the **Custom Model Views | 2-Import PDF** named view.

3. Using the **Insert** ribbon and the **Import** panel, select the **PDF Import** command. 🖹

4. Select the `Sample Project.pdf` file and click the **Open** button.

5. Using the **Import PDF** dialog, review the pages available to import. This PDF contains 35 pages, and you can select just the page you need. We will select **page 23**, the Expanded Plans page.

6. After selecting page 23, turn ON the **Specify insertion point on-screen** setting.

7. Set the **Scale** to **48**, as these Enlarged Plans are defined to plot at **1/4" = 1'-0"**.

8. In the **PDF data to import** section, define the following settings:

    - Vector geometry – **ON**, will convert vectors to AutoCAD objects

    - Solid fills   – **ON**, will convert solid fills to hatches

    - TrueType text   – **ON**, will convert TTF to text

    - Raster Images   – **ON**, will convert images to a PNG

> **Note**
>
> By default, AutoCAD SHX fonts will convert to geometry not text. I will demonstrate how to fix this later in this exercise.

9. In the **Layers** section, define the following setting:

    - Use PDF layers – **ON**, will convert layers from PDF to AutoCAD layers

10. In the **Import options** section, define the following settings:

    - Join line and arcs segments   – **ON**, will consolidate linework

    - Convert solid fills to hatches   – **ON**, will convert fills to hatches

    - Apply lineweight properties   – **ON**, will apply lineweights

    - Infer linetypes from colinear dashes – **ON**, will try to convert dashed linework into AutoCAD linetypes

11. Notice **PDF scale** is set to **1:1** for page 23. If the PDF was created using a specific scale, it would be displayed here. For example, select **page 13**, and the scale is listed as **1/8"=1'-0"**.

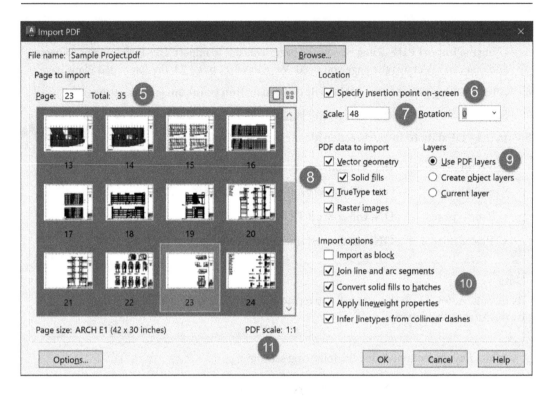

Figure 10.18: Import PDF dialog

12. Reselect **page 23**.

13. Click **OK** to save our changes and close the dialog.

14. *Left-click* in the drawing view to place the PDF-converted objects.

15. **Zoom In** to the LOWER RIGHT corner of the border file and review the text conversions. The TTF text objects (ENLARGED BATHROOM PLAN and A402) are still editable text objects, but the SHX text objects (2008.0 and A402 ENLARGED BATHROOM PLANS.DWG) are just linework and cannot be easily edited.

16. Using the **Annotate** ribbon and the **Text** panel, select the dialog launcher icon to open the Text Style dialog.

Prior to importing the PDF, the only text style in this drawing file was the STANDARD text style. After importing the PDF, the TTF text was converted to text using a PDF prefix style name.

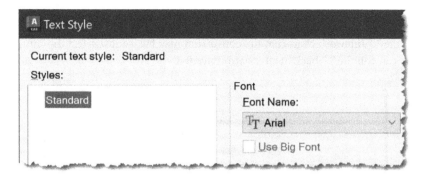

Figure 10.19: Initial Text Styles

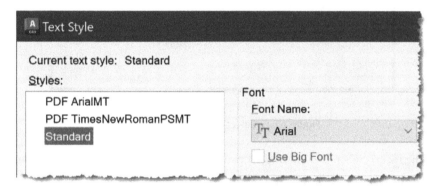

Figure 10.20: PDF Import text style results

Converting Linework Text to Mtext

We can use the RECOGNIZE SHX TEXT command to convert the linework text to MTEXT, and AutoCAD will attempt to convert it to a matching SHX font/text style.

The RECOGNIZE SHX TEXT command is a post-processing tool for converting selected geometry into single-line MTEXT objects on the current layer. Be careful to select only the SHX linework text.

First, we need to select all the linework text.

1. Select the large **A** linework character in the titleblock and *right-click* to access the **Properties** dialog. Note that the layer for this linework text is **PDF_A-Anno-Titl**, which was created during the **Import PDF** process.

2. Using the **Home** ribbon and the **Layers** panel, select the **Make Current** command ![icon] to set this layer as the current layer. This will preserve the layer standards from the PDF during this conversion process.

3. Reselect the large **A** linework text again and *right-click* to access the **Select Similar** command. This will select all the linework text on the **PDF_A-Anno-Titl** layer.

> **Note**
>
> If you select the "." linework character, the conversion may fail, so just select the "numbers" graphics. We can edit the "**.**" back in afterward if needed.

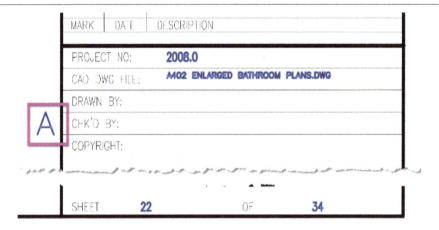

Figure 10.21: Select Similar results 1

4. Using the **Insert** ribbon and the **Import** panel, select the **Recognize SHX Text** command.

5. Click **Close** to close the **Recognize SHX Text** dialog.

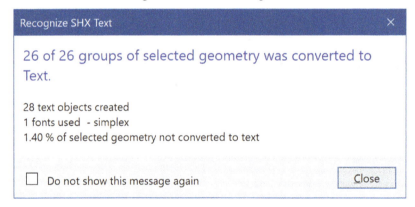

Figure 10.22: RECOGNIZE SHX TEXT results 1

6. Select one of the LINEAR linework text characters and *right-click* to access the **Select Similar** command. This will select all the LINEAR characters. Note that the layer for this linework text is **PDF_A-Anno-Titl-Note**.

We can avoid having to change our current layer using the text **Recognition Settings**:

1. Using the **Insert** ribbon and the **Import** panel, select the **Recognition Settings** command.

2. Using the **PDF Text Recognition Settings** dialog, modify the **Create Text on** setting to use **Same Layer as Geometry**. Click **OK** to close the dialog.

3. Reselect the LINEAR linework text again and *right-click* to access the **Select Similar** command. This will select all the LINEAR characters.

4. Select one of the CURVED linework text characters and *right-click* to access the **Select Similar** command. This will select all the CURVED characters.

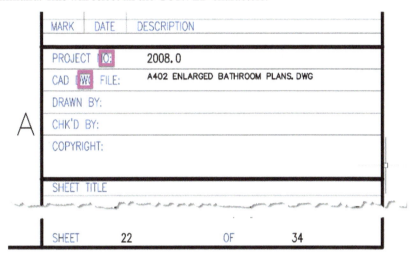

Figure 10.23: Select Similar results 2

It might take a few more selections to get all the linework text, depending on the layer results from the PDF, but keep using the SELECT SIMILAR command to select linework text as needed:

1. Using the **Insert** ribbon and the **Import** panel, select the **Recognize SHX Text** command.

2. Click **Close** to close the **Recognize SHX Text** dialog.

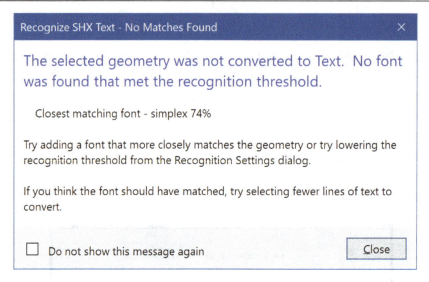

Figure 10.24: RECOGNIZE SHX TEXT ERROR results

More often than not, you will get an error if special characters are selected. Let me demonstrate how to fix this:

1. In this example, the ":" and " ' " characters are causing this error. After selecting all the characters, use any method you prefer to de-select the ":" and " ' " characters from the selection set prior to running the RECOGNIZE SHX TEXT command.

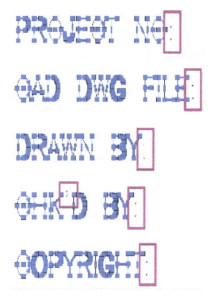

Figure 10.25: Special Characters De-Selected

2. Using the **Insert** ribbon and the **Import** panel, select the **Recognize SHX Text** command.

3. Click **Close** to close the **Recognize SHX Text** dialog.

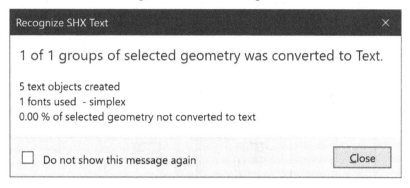

Figure 10.26: RECOGNIZE SHX TEXT results 2

Yes, this process can be tedious, but what a great way to convert PDF data to DWG data, right?

Recognition Settings

Now, let's look at some of the other settings that control the SHX text conversion:

1. Continue using the `10-4_Working with PDF Files.dwg` file.

2. Using the **Insert** ribbon and the **Import** panel, select the **Recognition Settings** command.

3. Using the **PDF Text Recognition Settings** dialog, review the fonts listed. If you use custom fonts, you will want to add them to this list for better text recognition.

4. Click the **Add** button and navigate to add the following font file for this exercise:

   `..\ACAD_TipsTechniques\Exercise_Files\Company XYZ\support\ ARCHITXT.SHX`.

5. Click **Open** to add this custom font to the list and click **OK** to close this dialog.

6. Select the ARCHITECTURAL linework text and *right-click* to access the **Select Similar** command.

7. Using the **Insert** ribbon and the **Import** panel, select the **Recognize SHX Text** command.

8. Click **Close** to close the **Recognize SHX Text** dialog.

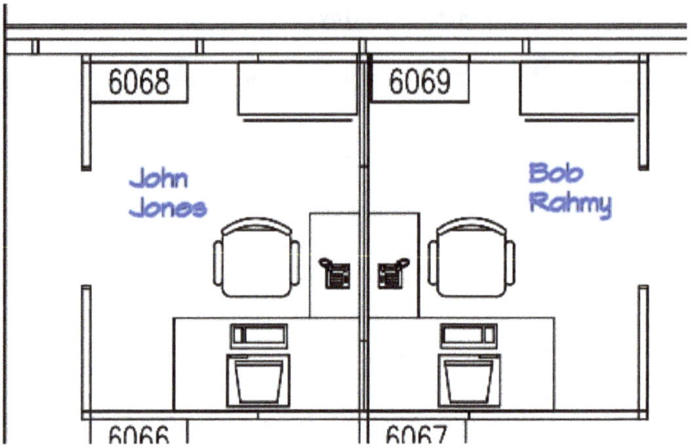

Figure 10.27: PDF Text Recognition results 3

> **Note**
>
> You might need to increase the recognition threshold in the PDF TEXT RECOGNITION SETTINGS dialog to around 85% the first time you try to recognize a custom font, but after the first successful recognition, you can increase it to 95% again.

Using Bonus Commands

In this section, we will look at how to use the REMOTE TEXT feature to extract data from external files, similar to reference files, and use it to document the contents of a file or print.

All of you should know that you can print with the out-of-the-box plot stamp provided in the PRINT dialog. This plot stamp provides you with the filename and date information. This can be somewhat limiting if you need to know what files are contained in the printed output, for example, what references are displayed in this print. This is where REMOTE TEXT can provide you with customizable reference information as part of your plot stamp.

Remote Text (Express Tools)

Remote text is used to display text from an externally linked file, such as a sheet note, a legal disclaimer, or a sheet stamp, which is common to several drawings. You can also use it to display larger bodies of text, such as specifications or assembly instructions, that are frequently updated throughout the design process.

Remote text objects are displayed the same as AutoCAD TEXT and MTEXT objects, but the source for the text content is either an external text file or the value of a DIESEL expression, which is demonstrated in the following exercises.

REMOTE TEXT	Command Locations
Ribbon	Express Tools	Text (expanded)	Remote Text
Command Line	RTEXT (RTEX)

In the first example, we will look at how to use RTEXT to display a disclaimer block on the drawing.

RTEXTAPP	
Defines what application is used to edit REMOTE TEXT objects.	
Type: Integer	
Saved in: Drawing	
0 (default)	Off
1	Detects when new layers have been to attached references
2	Detects when new layers have been added to the drawing and references

Remote Text Object

In this exercise, we will learn how to place a REMOTE TEXT object.

1. Open the 10-5_Reference BONUS Commands.dwg file.

2. Using the **View** ribbon and the **Named Views** panel, restore the **Custom Model Views | 1-FILE Remote Tex**t named view, using the **Restore View** *drop-down list*.

3. Using the **Express Tools** ribbon and the **Text** expanded panel, select the **Remote Text** command. 🄰

4. Use the *Enter* key to accept the **File** command option and navigate to the file, and click **Open** to close the dialog and open the following file.

    ...\ACAD_TipsTechniques\Exercise_Files\Chapter 10\refs\Submittal Stage-Preliminary.txt

5.   *Left-Click* above the existing **Preliminary Stamp BLOCK** object to place the REMOTE TEXT contents.

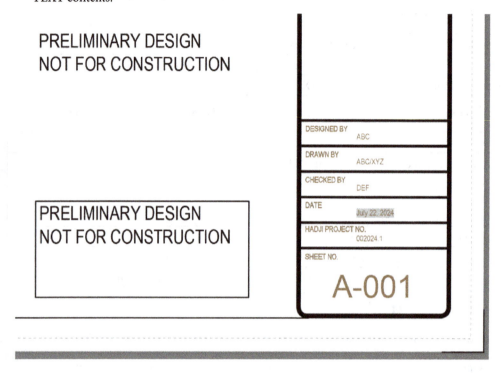

Figure 10.28: REMOTE TEXT results 1

This REMOTE TEXT is not editable using the typical AutoCAD text edit commands; you must use the REMOTE TEXT command to make changes in the external file that is linked to this text object. You can also use the PROPERTIES dialog to make changes to the content or appearance of the REMOTE TEXT object.

In this exercise, we will learn how to use and place a STAMP block object that will display the text from an external text file, making it a generic STAMP block. Use this type of block for submittal stages, construction notes, or anything that changes throughout the design process.

1.   Continue using the `10-5_Reference BONUS Commands.dwg` file.

2.   *Double-left-click* to edit the REMOTE TEXT object and note that you cannot edit this RTEXT object as you would a normal TEXT object.

3.   Using Windows Explorer, navigate to the linked external text file and open the file in your favorite text editor application:

`...\ACAD_TipsTechniques\Exercise_Files\Chapter 10\refs\Submittal Stage-Preliminary.txt`

4. Edit the contents of the text file to include the additional line of text:

    ```
    SUBMITTAL STAGE: 30%
    ```

5. Close the text file and save these changes.

6. Return to AutoCAD and using the Command Line, key in the REGEN command to update the current drawing file.

7. The REMOTE TEXT object updates to include the new addition from the external text file.

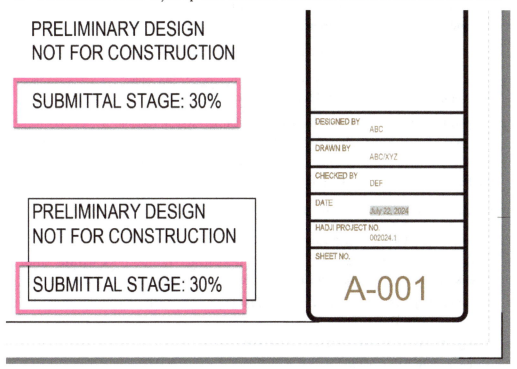

Figure 10.29: REMOTE TEXT File results 2

8. The REMOTE TEXT capabilities are not intended to be used for data where ATTRIBUTES or FIELD objects are available, but for data objects that are commonly edited and are repeatedly used throughout the drawing or sheet sets.

> **Note**
> If a recipient of your drawings does not have the Express Tools installed, the REMOTE TEXT object will display as a "bounding box" and they will receive an error when opening the file. The fix for this would be to explode the REMOTE TEXT object before sending out the drawing file, which breaks the link to the external text file and converts the REMOTE TEXT object to a normal MTEXT object.

In the next exercise, we will learn how to use REMOTE TEXT DIESEL objects to extract data from our drawing. In this example, we will extract a list of attached reference files and place this list in our border file.

Remote Text Diesel Expression

So, what is a DIESEL expression in AutoCAD?

A DIESEL (Direct Interpretively Evaluated String Expression Language) expression evaluates an input text string and generates output text string results.

The strings generated by DIESEL expressions can be used to do the following:

- Control the labels and behavior of buttons on ribbon panels and toolbars in Windows-based releases of AutoCAD and AutoCAD LT products

- Display text on the status bar with the MODEMACRO system variable in Windows-based releases of AutoCAD and AutoCAD LT products

- Manipulate string values with the AutoLISP menucmd function; AutoLISP is not supported in the AutoCAD LT for MacOS product

In this exercise, we will learn how to use a DIESEL expression to list the attached reference files.

1. Continue using the 10-5_Reference BONUS Commands.dwg file.

2. Using the **View** ribbon and the **Named Views** panel, restore the **Custom Model Views | 2-DIESEL Remote Text** named view, using the **Restore View** *drop-down list*.

3. Using the **Express Tools** ribbon and the **Text** expanded panel, select the **Remote Text** command. Ⓐ

4. Using the Command Line, select the **Diesel** command option.

5. Using the **Edit Rtext** dialog, key in $ (xrefs,2) and click **OK** to close the dialog.

6. *Left-click* in the **Reference Files** area of the border to place the list of attached reference files.

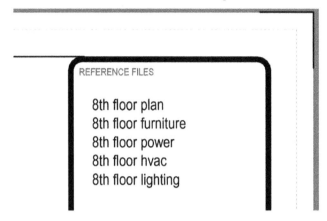

REFERENCE FILES

8th floor plan
8th floor furniture
8th floor power
8th floor hvac
8th floor lighting

Figure 10.30: DIESEL Remote Text Results

This list will update automatically as references are added and removed to the sheet layout. Add this REMOTE TEXT object to your standard border to generate a reference list for all drawing files in your layouts.

So, what does the DIESEL text $(xrefs,2) mean?

This XREFS is not a command in AutoCAD but a function of the Express Tools REMOTE TEXT. The following displays the results for the various options available using the XREFS command:

```
$(xrefs [, flags [, leader [, trailer]]])
```

XREFS Parameters

FLAGS	
Flags are a collection of bitcode numbers, each with the following meanings:	
1 (default)	Includes the Xref filename (not exclusive to flag 2)
2	Includes the Xref block name (not exclusive to flag 1)
4	Do not display the filename extension
8	Do not display the file path
16	Show nesting with additional spacing

For example, $(xrefs,1) displays both the filename and the relative path.

.\refs\8th floor plan.dwg
.\refs\8th floor furniture.dwg
.\refs\8th floor power.dwg
.\refs\8th floor hvac.dwg
.\refs\8th floor lighting.dwg

Figure 10.31: $(xrefs,1) Results

For example, $(xrefs,3) displays both the block name and the filename with the relative path.

8th floor plan [.\refs\8th floor plan.dwg]
8th floor furniture [.\refs\8th floor furniture.dwg]
8th floor power [.\refs\8th floor power.dwg]
8th floor hvac [.\refs\8th floor hvac.dwg]
8th floor lighting [.\refs\8th floor lighting.dwg]

Figure 10.32: $(xrefs,3) Results

For example, $(xrefs,4) displays both the filename and relative path, but without the file extension.

.\refs\8th floor plan
.\refs\8th floor furniture
.\refs\8th floor power
.\refs\8th floor hvac
.\refs\8th floor lighting

Figure 10.33: $(xrefs,4) Results

For example, $(xrefs,8) displays the filename with the file extension.

8th floor plan.dwg
8th floor furniture.dwg
8th floor power.dwg
8th floor hvac.dwg
8th floor lighting.dwg

Figure 10.34: $(xrefs,8) Results

Leader

A LEADER is a text string that will be inserted before each Xref entry.

For example, $(xrefs,2, Includes:) displays the block name or filename with the Includes: prefix.

Includes: 8th floor plan
Includes: 8th floor furniture
Includes: 8th floor power
Includes: 8th floor hvac
Includes: 8th floor lighting

Figure 10.35: $(xrefs,2,Includes:) results

Trailer

A TRAILER is a text string appended to each Xref entry except the last.

For example, $(xrefs,2, ,) displays the filenames in a horizontal mode.

8th floor plan 8th floor furniture 8th floor power 8th floor hvac 8th floor lighting

Figure 10.36: $(xrefs,2, ,) results

Use the following link to review the reference help available for AutoCAD Express Tools:

https://help.autodesk.com/view/ACD/2016/ENU/?guid=GUID-CC626232-DC3A-45E1-B3C8-DF3F79186DE2

Many of the REMOTE TEXT commands have been replaced using the new FIELD objects in AutoCAD. However, some of these old features are still only available using the old REMOTE TEXT features, such as the reference fileS command option demonstrated here. Check out what capabilities are available using REMOTE text; you will be amazed at what is possible!

Summary

In this chapter, we learned how to work with reference files and PDF files as references and imports. We learned how to improve your work environment when working the reference files and how to take advantage of PDF data when that is all that is available. Use these techniques to improve your data sharing with users inside and outside your organization.

In the next chapter, we will look at how to automate CAD Standards using Tool Palettes.

11

Enforcing Your CAD Standards

This chapter will not teach you how to define your CAD Standards but, rather, how to enforce the standards in your project files. I will assume you already know the importance of defining the standard layers, linetypes, text styles, dimstyles, and other resources to maintain corporate drawing standards, not individual drawing standards. Our goal is to achieve drawing consistency between individual users. Whether your organization has 2 or 2,000+ users, this consistency is key to producing a corporate look and feel in your project data. As a user in any organization, you should be able to work in any drawing and expect certain rules to apply.

In this chapter, we'll cover the following topics:

- Setting up CAD Standard resources
- Configuring CAD Standards checker
- Using CAD Standards checker
- Using Layer Translator

By the end of this chapter, you will be able to make your entire team more productive and efficient by using and managing your organization's CAD Standards. You will learn how to fix inconsistencies in your drawing files to create standardized content in AutoCAD.

Setting up CAD Standard resources

I think the Standards Checker feature is one of the most overlooked features in AutoCAD.

First, we must locate our CAD Standard resources and tell AutoCAD where to find them. Although every organization distributes its standards file differently, try to standardize this structure and naming convention as much as possible.

Working with CAD Standards files

For this exercise, I have delivered predefined standards files (.DWS) in the following location:

`..\ACAD_TipsTechniques\Exercise_Files\CompanyXYZ\standards\XYZ_Standards.dws`

You can choose to include all CAD Standards resources in a single file or multiple files, based on their content. My personal preference is the multiple file configuration, as it is easier to maintain and update.

Global Standards file

The following file contains all global resources.

XYZ_Standards.dws	Contains standards for all content specific resources

Content-Specific Standards files

The following files are broken down to contain discipline-specific resources:

XYZ_Layers.dws	Contains standard layers
XYZ_Tables.dws	Contains standard table styles
XYZ_Linetypes.dws	Contains standard linetypes
XYZ_Annotation.dws	Contains standard text styles, dim styles, and leader styles
XYZ_Civil.dws	Contains CIVIL discipline styles
XYZ_Architectural.dws	Contains ARCHITECTURAL discipline styles

Depending upon the amount of defined resources, you will probably need to break these files down even more, but for this course, this will give you a good representation of the concepts.

Let's look at what is stored in these files:

1. Open the `11-1_Working with CAD Standards.dwg` file.

2. Open the `...\ACAD_TipsTechniques\Exercise_Files\CompanyXYZ\standards\XYZ_Layers.dws` file.

   This file contains the standard layer definitions, which include some linetypes. Always be sure to assign only your approved linetype CAD Standards to these layers. Create your standard LAYER FILTERS in this file.

3. Using the **Home** ribbon and the **Layers** panel, select the **Layer Manager Properties** command and review the defined layers.

4. Close the `..\ACAD_TipsTechniques\Exercise_Files\CompanyXYZ\standards\XYZ_Layers.dws` file.

5. Continue using the `11-1_Working with CAD Standards.dwg` file.

6. Using the **Manage** ribbon and the **CAD Standards** panel, select the **Configure** command.

> **Note**
>
> If you select the CHECK command and the file is not configured with CAD Standards, an error will display, alerting you to configure the resources to check against. The Check Standards will then run automatically.

7. Using the **Configure Standards** dialog, click the **Add Standards File** button ➕ and select the `..\ACAD_TipsTechniques\Exercise_Files\XYZ Standards\XYZ_Layers.dws` file.

8. Select the **Plug-ins** tab and *right-click* at the TOP of the list to access the **Clear All** command. Then, place a checkmark for the **Layers** plug-in and click **OK** to close the dialog.

9. Using the **Layer Manager Properties** dialog, click the **New Layer** button, 🗂, and notice the Standards Violation notification in the LOWER-RIGHT corner of the application.

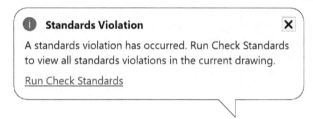

Figure 11.1: Layer Standards notification

Because **Layer1** does not exist in the `XYZ_Layers.dws` file, the new layer is flagged as non-compliant. It's great that it keeps us from creating non-compliant layers, but this is not user-friendly when you are trying to work in a file. Fortunately, we can change this behavior. Let's change the default notification settings so that we aren't notified about every non-compliant standards violation and can run the Standards Checker process at our convenience:

1. Using the **Layer Manager Properties** dialog, click the **Delete Layer** button 🗂 to delete the **Layer1** layer.

2. Using the **Manage** ribbon and the **CAD Standards** panel, select the **Configure** command.

3. Using the **Configure Standards** dialog, click the **Settings** button.

4. Using the **CAD Standards Settings** dialog, select the **Disable Standards Notifications**.

Figure 11.2: CAD Standards settings

You can also automatically fix non-standard properties. However, I will caution you about this, as it does not give you a chance to review the non-compliant issue. Maybe after you have used the CAD Standards Checker and are confident that your Standards are thoroughly defined, you can turn this setting on.

Adding additional CAD Standard resources

Next, we will add additional CAD Standards resource files for additional checking:

1. Continue using the 11-1_Working with CAD Standards.dwg file.

2. Using the **Manage** ribbon and the **CAD Standards** panel, select the **Configure** command.

3. Using the **Configure Standards** dialog, click the **Add Standards File** button ➕ and select these additional files:

   ...\ACAD_TipsTechniques\Exercise_Files\CompanyXYZ\standards\ XYZ_Annotation.dws and the XYZ_Linetypes.dws

4. Select the **Plug-ins** tab and add a checkmark for the following plug-ins.

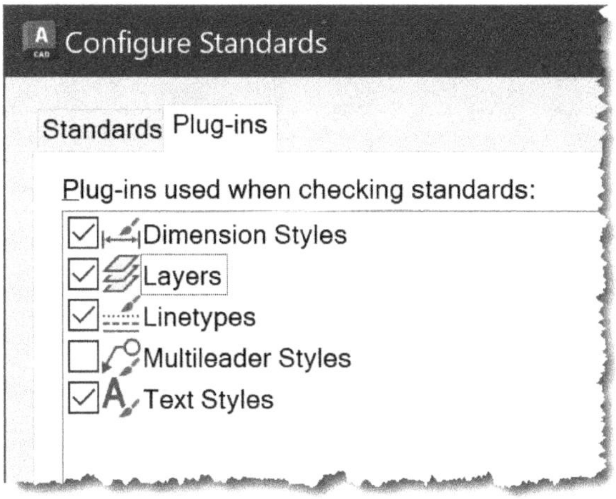

Figure 11.3: CAD Standards plug-ins

5. Click the **Check Standards** button to check this file again, using all the associated resources.

Using the **Check Standards** dialog, you can review the problems and select fixes.

The first problem is with the Standard dimstyle:

1. Select the **XYZ_Standard** dimstyle from the XYZ_Annotation.dws file, and note the differences between the **Standard** dimstyle and the **XYZ_Standard** dimstyles in the LOWER portion of the dialog.

2. Click the **Fix** button to make the necessary changes. This actually replaces the Standard dimstyle in the current file with the **XYZ_Standard** dimstyle from the .DWS file.

3. Click the **Close** button to close the **Check Standards-Check Complete** dialog.

4. Click the **Next** button, and the **P-FIXT** layer name is flagged as non-compliant. In this instance, use the **Mark this problem as ignored** option. We can address this layer later as needed.

5. Click **Next**, and you will be notified that the S-COLS layer is the wrong color in this file. The S-COLS layer should be GREEN, not BLUE. Click on **Fix** to change the layer color back to GREEN.

6. Click **Next**, and you will be notified that the **Layer10** layer is flagged as non-compliant. Select the **A-DEMO** layer, and the contents of **Layer10** will then merge with the **A-DEMO** layer.

7.   When the standards checking process is complete, you will be presented with the **Check Standards** dialog that displays the number of problems found, fixed, and ignored.

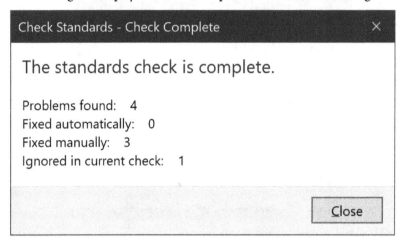

Figure 11.4: Check Standards results

8.   Click the **Close** button to close the **Check Standards - Check Complete** dialog, and click on **Close** again to close the **Check Standards** dialog.

> **Note**
> If you ignore a problem during the standards checking process, the problem is flagged with your name.

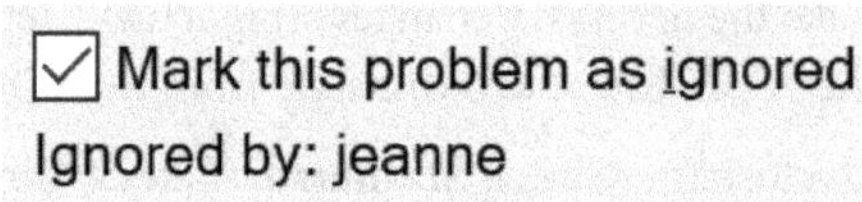

☑ Mark this problem as ignored
Ignored by: jeanne

Figure 11.5: An ignored Standards Checking issue

Restoring Layer Standards

Be sure to use a Layer State to save your default layer standards. Since Layer States can be imported into any project drawing file, they can be used to restore layer standards easily.

In this example, we have a project file with purged layers and modified colors and lineweights by a user:

1. Open the `11-2_Project 1.dwg` file.

2. Using the **Home** ribbon and the **Layers** panel, select the **Layer Manager Properties** command.

3. Using the **Layer Manager Properties** dialog, click the **Layer States Manager** button.

4. Using the **Layer States Manager** dialog, click the **Import** button and navigate to the layer standards `XYZ_Layers.dws` file.

5. Select the **XYZ_Layer Standards** layer state and click **OK** to complete.

6. Select the **XYZ_Layer Standards** layer state, and click the **Restore** button to restore the standard layer conditions to the current drawing file.

Magically, your layer standards are easily restored! Well, not magically, but with a little planning and well-defined standards files, you, too, can fix most CAD Standards layer issues.

Using the LAYER STATES MANAGER dialog, you can also EXPORT your Layer States to a common location so that they are available for import. Don't forget to re-export and update your Layer State after making changes to the standard layers. This is not dynamic, and you have to manually maintain these resources. A Layer State is a snapshot in time of the layer properties. Review Layer States in *Chapter 9*.

Batch Standards Checker

You can batch-check your files for compliant CAD Standards. Using the Batch Standards Checker, you can use the associated standards (`.DWS`) files or check the standards with a new selection of standards (`.DWS`) files. The Batch Standards Checker also allows you to check the attached reference files:

1. Using **Windows Start | AutoCAD 2025**, select the **Batch Standards Checker** utility.

2. Using the **Batch Standards Checker** dialog, start with the **Drawings** tab, click the **Add Drawings** button , and select the following files:

    `..\ACAD_TipsTechniques\Exercise_Files\Chapter 11\11-2_Project 2.dwg` and `11-2_Project 3.dwg`

3. Select the **Standards** tab, and select the option to **Check all drawings using the following standards files**.

4. Click the **Add Standards** button and select the following files:

    `XYZ_Layers.dws, XYZ_Linetypes.dws` and `XYZ_Annotation.dws`

5.   Select the **Plug-ins** tab and verify that all plug-ins are selected and enabled as needed.

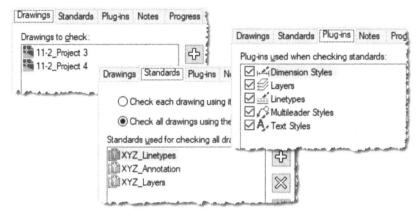

Figure 11.6: Batch Standards Checker

6.   Continue using the **Batch Standards Checker** dialog, and click the **Save** button 🖫 to save these settings to a Standards Checker file (. CHX).

7.   Click the **Start Check** button 🏃 to run the Standards Checker on the selected files.

8.   The **STANDARDS AUDIT REPORT** will open automatically to display the Standards Checker results.

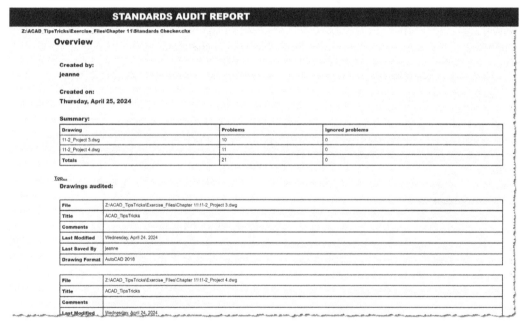

Figure 11.7: Batch Standards Checker results

Use the Batch Standards Checker to check any drawings coming in from a consultant or sub-contractor to verify that they followed the required CAD Standards.

In the next section, we will learn how to use the Layer Translator to fix the layers of a non-standard drawing file to comply with our current or new CAD Standards.

Using the Layer Translator

The Layer Translator provides an efficient method for fixing layers in older drawing files or files from outside your organization. This tool provides an easy interface for selecting layers and mapping them to alternate layer definitions:

1. Open the `11-3_Using Layer Translator 1.dwg` file.

2. Using the **Manage** ribbon and the **CAD Standards** panel, select the **Layer Translator** command.

    The first thing you need to do is select a layer mapping file that contains the current layer standards. We will use the **XYZ_Layers** file we used earlier in this chapter:

3. Using the **Layer Translator** dialog, click the **Translate to | Load** button and select the `XYZ_Layers.dws` file. This loads the standard layer definitions for the mapping process.

    Now, you have the current file layers in the TRANSLATE FROM list and the standard layers in the TRANSLATE TO list. Next, map all the layers with the same name in both lists to eliminate them from the manual selection process.

4. Click the **Map Same** button and review the list of layers displayed in the **Layer Translation Mapping** section.

Layer Translation Mappings

Old Layer Name	New Layer Name	Color	Linetype	Lineweight	Transparency	Plot style
0	0	7	Continuous	0.00 mm	0	ByColor
Defpoints	Defpoints	7	Continuous	0.00 mm	0	ByColor
0-XREFS	0-XREFS	255	Continuous	0.00 mm	0	ByColor

Figure 11.8: Initial Layer Translation mappings

5. Review the mapping criteria and click the **Save** button to save these mappings to the `11-3_Layer Mappings.dws` file for future use.

6.   Using the **Translate From** list, select the following layers and map them to the **Translate To** list:

Translate From	Translate To
COLUMNS	S-COLS
DOORS	A-DOOR
ELEVATORS	A-FLOR-ELEV
FIXTURES	P-FIXTURE
MECHANICAL	M-HVAC-EQPT
NOTES	A-ANNO-NOTE
PARTITIONS	P-EQPT-PART
ROOM NAMES	A-ANNO-RMNAME
STAIRS	A-FLOR-STRS
THERMAL	M-CONT-THER
WALLS-EXT	A-WALL-EXT
WALLS-INT	A-WALL-INT
WINDOWS	A-GLAZ

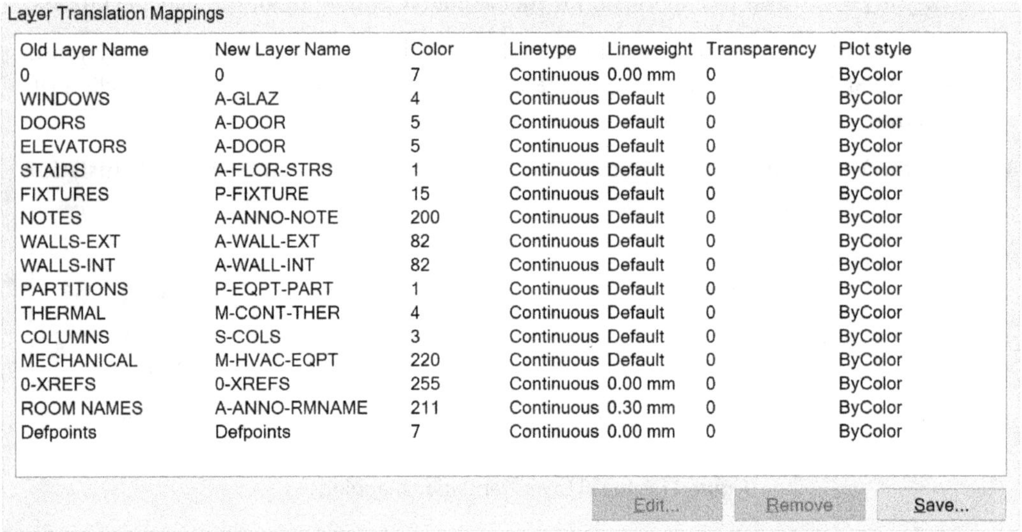

Layer Translation Mappings

Old Layer Name	New Layer Name	Color	Linetype	Lineweight	Transparency	Plot style
0	0	7	Continuous	0.00 mm	0	ByColor
WINDOWS	A-GLAZ	4	Continuous	Default	0	ByColor
DOORS	A-DOOR	5	Continuous	Default	0	ByColor
ELEVATORS	A-DOOR	5	Continuous	Default	0	ByColor
STAIRS	A-FLOR-STRS	1	Continuous	Default	0	ByColor
FIXTURES	P-FIXTURE	15	Continuous	Default	0	ByColor
NOTES	A-ANNO-NOTE	200	Continuous	Default	0	ByColor
WALLS-EXT	A-WALL-EXT	82	Continuous	Default	0	ByColor
WALLS-INT	A-WALL-INT	82	Continuous	Default	0	ByColor
PARTITIONS	P-EQPT-PART	1	Continuous	Default	0	ByColor
THERMAL	M-CONT-THER	4	Continuous	Default	0	ByColor
COLUMNS	S-COLS	3	Continuous	Default	0	ByColor
MECHANICAL	M-HVAC-EQPT	220	Continuous	Default	0	ByColor
0-XREFS	0-XREFS	255	Continuous	0.00 mm	0	ByColor
ROOM NAMES	A-ANNO-RMNAME	211	Continuous	0.30 mm	0	ByColor
Defpoints	Defpoints	7	Continuous	0.00 mm	0	ByColor

Edit...   Remove   Save...

Figure 11.9: Layer Translation mappings 2

1. Click the **Translate** button and save these mappings to the `11-3_Layer Mappings.dws` file for future use.

2. Open the `11-3_Using Layer Translator 2.dwg` file.

3. Review the layers using the **Layer Manager Properties** dialog.

4. Using the **Manage** ribbon and the **CAD Standards** panel, select the **Layer Translator** command.

5. Again, load the `11-3_Layer Mappings.dws` file, and this time, we will click on the **Edit** button to change the mapping color for this layer.

6. Using the **Edit Layer** dialog, change the color from **211** to **Magenta**, and then click **OK** to close the dialog.

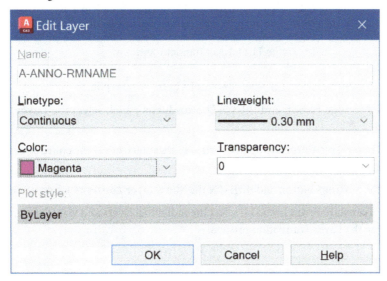

Figure 11.10: Layer mapping settings

7. Click the **Translate** option to translate the layers to the new layer definitions.

When using the LAYER TRANSLATOR command, you can control the translation settings using the following SETTINGS dialog. One setting that is turned off by default is the ability to see visually what drawing contents are on the layer to be translated.

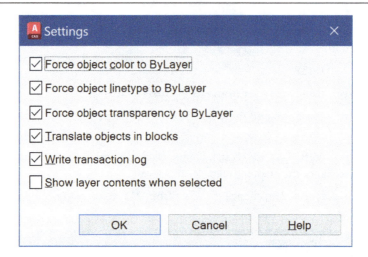

Figure 11.11: Layer Translator settings

8.   Open the 11-3_Using Layer Translator 3.dwg file.

9.   Using the **Manage** ribbon and the **CAD Standards** panel, select the **Layer Translator** command.

10. Move the **Layer Translator** dialog in the view so that you can see the entire floor plan outside the dialog.

11. Click on the **Settings** button and turn ON the **Show layer contents when selected** setting.

12. Using the **Translate From** layer list, select the individual layers and view the contents of each layer during the Layer Translation process.

Figure 11.12: STAIRS preview

Figure 11.13: ROOM NAMES Preview

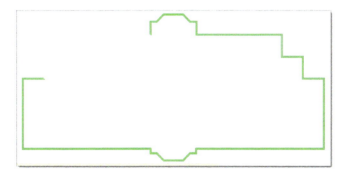

Figure 11.14: WALLS-EXT results

Summary

In this chapter, we learned how to enforce CAD Standards within our project drawing files to standardize drawings between various users. Maintaining this consistency is critical to the look and feel of any organization's data files. Learning to apply these standards during the drafting process or afterward is a concept that can be used within your specific workflow.

In the next chapter, we will examine how to make the most of those common commands that you use every day. Do you know all the available command options? Let's find out!

12

Making the Most of Common Commands

This chapter will teach you how to use the most commonly used commands differently, such as TRIM, PEDIT, and ARRAY. You will also learn how to change common commands and make them repeat automatically by using the MULTIPLE command. After improving your use of common commands, I will introduce you to some hidden commands that many users don't discover during their everyday use of AutoCAD.

In this chapter, we'll cover the following topics:

- Using object commands
- Using the hidden commands
- More object snaps
- Using the calculators
- More bonus commands

By the end of this chapter, you will be able to improve your use of these everyday common commands, as well as understand how to use the built-in calculators to control your linetype scales.

Using object commands

In this section, we will examine how to use specific commands to modify objects in a drawing, such as the REVERSE command, how to automatically repeat commands, and how to use the WIPEOUT and SUPERHATCH commands. We will also investigate hidden features in the ARRAY, TRIM, ARC, and POLYLINE commands.

TRIM Objects

We all use the TRIM command daily, but did you know you can trim more than linear objects? Of course, you can!

Did you know you can trim a HATCH object? The next example demonstrates using the TRIM command to trim a HATCH object:

1. Open the `12-1_Using OBJECT Commands.dwg` file.

2. Using the In-Canvas View Controls, restore the **Custom Model Views | 1-Trim Hatch** named view.

    First, the HATCH object must have a BOUNDARY associated with it to be trimmable.

3. Using the **Layer** *drop-down list*, THAW the **1-Trim Objects** layer.

4. Select the ceiling grid HATCH object and *right-click* to access the **Set Boundary** command.

5. Using the **Home** ribbon and the **Modify** panel, and select the **Trim** command.

6. Select the HATCH object outside the BLUE POLYLINE shape, and the HATCH object is trimmed to the POLYLINE.

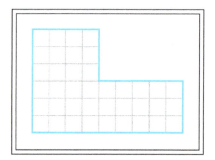

Figure 12.1: The Trim HATCH result

In the next exercise, we will learn how to use WIPEOUT objects to reduce modifications.

Using WIPEOUT

In this example, we will use a WIPEOUT object in a light fixture block to eliminate the need to trim the ceiling grid hatch object. First, let's look at the Light Fixture-2x4 block. For this block, I created a WIPEOUT object as a background so that, when placed, it will "hide" the ceiling grid linework, and I no longer need to "trim" the linework:

1. Continue using the `12-1_Using OBJECT Commands.dwg` file.

2. Using the In-Canvas View Controls, restore the **Custom Model Views | 2-Wipeouts** named view.

Before we add the Light Fixture to the ceiling grid, we must change the AutoCAD setting for the OSNAPHATCH system variable

OSNAPHATCH	
Controls whether object snaps will snap to a hatch pattern.	
Type: String	
Saved in: Drawing	
0 (default)	Ignore hatch objects when snapping
1	Recognize hatch objects when snapping

1. Using the Command Line, key in OSNAPHATCH and then key in the value of 1.

2. Select the **Light Fixture-2x4** block and *right-click* to access the **Add Selected** command.

3. Using the INTERSECTION snap, place the **Light Fixture-2x4** block in the ceiling grid where needed.

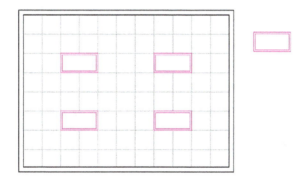

Figure 12.2: Using WIPEOUT's results

4. Using the GRIP, select one of the blocks overlapping the HATCH object and move it to a new location. The HATCH object is intact, so you can easily modify the design without having to repair it.

How many of your blocks can benefit from the addition of a WIPEOUT?

WIPEOUT	Command Locations
Ribbon	Home	Draw	Wipeout
	Annotate	Markup	Wipeout
Command Line	WIPEOUT (WI)

The WIPEOUT object has an outline that can be plotted, but most of us prefer it to be a non-plotting frame. Use the following system variables to control your WIPEOUT objects.

WIPEOUTFRAME	
Controls the display and plotting of wipeout frames.	
Type: Integer	
Saved in: Drawing	
0	Frames are not displayed or plotted. The frame will temporarily display during image selection.
1 (default)	Frames are displayed and plotted.
2	Frames are displayed but will not plot.

FRAMESELECTION	
Controls the display of the hidden frame for an image, underlay, reference, or wipeout objects.	
Type: Integer	
Saved in: Registry	
0	Hidden frames cannot be selected.
1 (default)	Hidden frames can be selected.

More POLYLINE Edits

We all use the EDIT POLYLINE (PEDIT) command regularly, but do you know everything about its editing capabilities? Let's find out.

PEDITACCEPT

Use the PEDITACCEPT system variable to simplify your polyline editing. When you select a non-polyline object, you are asked, "Do you want to turn it into one?" Have any of you ever answered NO to this question? Me neither. What a silly question! Let's change the default value for this system variable so that we are NEVER asked this question again:

1. Continue using the 12-1_Using OBJECT Commands.dwg file.
2. Using the In-Canvas View Controls, restore the **Custom Model Views | 3-Polylines** named view.

PEDITACCEPT	
Suppresses the prompt "Do you want it to turn into one?" When the prompt is suppressed, the selected object is automatically converted to a polyline.	
Type: String	
Saved in: Drawing	
0 (default)	The prompt to convert to a polyline is displayed.
1	The prompt to convert to a polyline is suppressed.

1. Using the Command Line, key in PEDITACCEPT and change it to a value of 1.

2. Using the **Home** ribbon and the **Modify** expanded panel to access the **EDIT POLYLINE** command.

3. Select one of the LINE objects in the CAPSULE, and note that the question has been suppressed.

4. Use the *Esc* key to cancel the command.

In the next exercise, we will learn how to use the "fuzz factor" available in EDIT POLYLINE (PEDIT) using the MULTIPLE EDIT POLYLINE (MPEDIT) command.

The FUZZ Factor

Have you ever tried to join several lines using the EDIT POLYLINE command, only to find that it fails due to "sloppy" drafting? Did you know that there is a "fuzz factor" you can apply to close those gaps in the lines? This option is only available through the MULTIPLE option, so try it out!

1. Continue using the 12-1_Using OBJECT Commands.dwg file.

2. Select the EDIT POLYLINE command again, and then *right-click* to access the **Multiple** command option.

3. We want to join the LINE objects, so select the three LINE objects and *right-click* to access the **Join** command option.

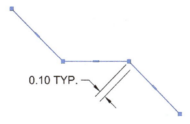

0.10 TYP.

Figure 12.3: PEDIT JOIN with EXTEND

The following figure displays the results with a vertex added to the polyline object.

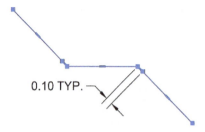

0.10 TYP.

Figure 12.4: PEDIT JOIN with ADD

4. Key in a "fuzz factor" distance of .2, which is larger than the gap in our lines, and press *Enter* to end the command.

All three lines are joined into a single POLYLINE object. The default JOINTYPE is set to EXTEND the lines to connect them. You can change the JOINTYPE to use EXTEND, ADD, or BOTH:

*   **Extend** will join the selected polylines by trimming or extending the line segments.

*   **Add** will join the selected polylines by adding a straight segment between the endpoints.

*   **Both** will join the selected polylines by trimming or extending first, adding a straight segment if needed.

> **Note**
>
> You can avoid having to select the MULTIPLE command option by using the MPEDIT command, which is the MULTIPLE EDIT POLYLINE command. This command is available as a key-in-only command.

> **Note**
>
> The JOIN command is an easier version of the PEDIT | JOIN function; however, it does not provide a "fuzz factor" command option.

Hopefully, this section demonstrated a few new uses for those everyday commands.

The Hidden Features of an Array

Many of us enjoy the intelligence available in the ARRAY command, but do you know everything about an associated ARRAY object?

ARRAY	Command Locations
Ribbon	Home	Modify	Array
Right-click Menu	Array
Command Line	ARRAY (ARRAY)

Deleting Items

Did you know you can delete individual items in an associated array? Let's try it out.

1.   Continue using the 12-1_Using OBJECT Commands.dwg file.

2.   Using the In-Canvas View Controls, restore the **Custom Model Views | 4-Hidden ARRAY** named view.

3.   Use the *Ctrl* key to select the two TREE objects inside the building.

4.   Use the *Delete* key on the keyboard to remove the selected items.

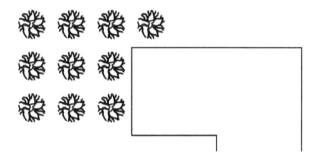

Figure 12.5: Removing Items in ARRAY

5.   Select the TREE ARRAY next to the building and *right-click* to access the **Array | Reset** command. Then, the ARRAY is reset to its original graphics, and the previously deleted items are restored.

Replacing Items in Array

Did you know you can replace individual items in an associated array?

1.   Continue using the 12-1_Using OBJECT Commands.dwg file.

2.   Select the TREE ARRAY object and *right-click* to access the **Array | Replace Item** command.

3.   Select the TREE–LIVE block as the replacement block and use the *Enter* key to continue.

4.   Press the *Enter* key to accept the **Centroid** base point for the new items.

5.   Select the three TREE–DEAD block items in the ARRAY to replace them with the TREE–LIVE block, as shown here.

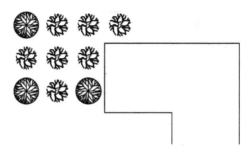

Figure 12.6: Replacing Items in ARRAY

Clip ARRAY

Treat the associated ARRAY object like a block, and you will soon realize that you can CLIP and NCOPY from an ARRAY, too. For the next example, we will create a dynamic offset of lines using the ARRAY command and boundary-clip the array. This will leave us with an editable offset of lines:

1. Continue using the 12-1_Using OBJECT Commands.dwg file.

2. Using the **Express Tools** ribbon and the **Modify** expanded panel to locate the **Extended Clip** command.

3. Select the POLYLINE shape as the CLIPPING FRAME, and select the horizontal lines ARRAY as the OBJECT TO CLIP.

4. Select the horizontal lines ARRAY, and using the **ARRAY Contextual** ribbon, change **ROWS| Distance Between** to 10.

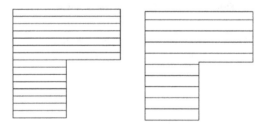

Figure 12.7: Clipping an ARRAY

Now, we have an easy-to-modify set of offset lines that can be adjusted as needed.

If your Express Tools ribbon tab is missing, try taking the following steps to access it. There are several causes for this missing tab:

- The Express Tools have not been enabled

- The Express Tools have not been installed

- The Express Tools installation is damaged

Not Enabled

Using the Command Line, key in EXPRESSTOOLS; the ribbon tab should automatically load. If not, try the next solution.

Damaged

If Express Tools have been installed, or if some Express Tools work but others do not, check the Support File Search Paths as follows:

1. Using the Command Line, key in the OPTIONS command to open the **Options** dialog and select the **Files** tab.

2. Expand the **Support Files Search Path** section and verify that the Express Tools path is present.

3. If the path is missing, use the **Add** button to add the following location:

    C:\Autodesk\Autocad 20xx\Express

Not Installed

Load the Express Tools .CUI file.

1. Use the **CUI** dialog to verify that the Express Tools are loaded as a **Partial Customization** file.

Deleting a Clip

The following steps can be used to delete an existing clip from a block, image, or reference:

1. Continue using the 12-1_Using OBJECT Commands.dwg file.

2. Using the **Insert** ribbon and the **References** panel, select the **Clip** command.

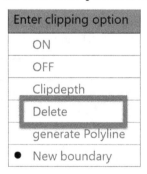

Figure 12.8: The Delete Clip command

This figure displays the DELETE CLIP command results.

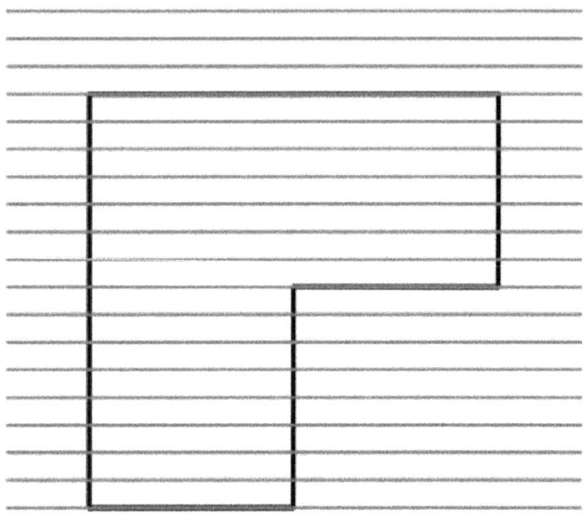

Figure 12.9: The Delete ARRAY Clip results

3. Select the ARRAY object we clipped in the previous exercise, and using the pop-up menu displayed, select the **Delete** command option.

In the next exercise, we will learn how to use the MULTIPLE command with common commands such as CIRCLE, RECTANGLE, OFFSET, and FILLET.

The MULTIPLE Command

Many commands in AutoCAD can be converted to MULTIPLE for repeated execution. For example, you can "multiple" or "repeat" the CIRCLE, PLINE, and many DIMENSION commands.

MULTIPLE	Command Locations
Command Line	MULTIPLE (MU)

Multiple CIRCLE

This exercise will demonstrate how to use the CIRCLE command with the MULTIPLE option:

1. Open the 12-1_Using OBJECT Commands.dwg file.
2. Using the In-Canvas View Controls, restore the **Custom Model Views | 5-MULTIPLE Commands** named view.
3. Using the Command Line, key in the MULTIPLE command and follow it by keying in the CIRCLE command, or use the C shortcut for the CIRCLE command.

4. *Left-click* in the drawing view to place the first circle. Note that where the CIRCLE command would normally end, the command continues running, allowing you to draw another circle. The command will run continuously until you use the *Esc* key to cancel it.

5. Next, let's try the MULTIPLE command for the RECTANGLE command.

Multiple RECTANGLE

This exercise will demonstrate how to use the RECTANGLE command with the MULTIPLE option.

1. Using the Command Line, key in the `MULTIPLE` command and follow it with the `RECTANGLE` command.

2. *Left-click* twice in the drawing view to place the first rectangle. Again, note that the RECTANGLE command continues running, allowing you to draw additional rectangles.

3. Use the *Esc* key to cancel the command when finished.

The next exercise will discuss how to use the OFFSET command to get multiple offsets without selecting lines multiple times. This command has a MULTIPLE option built into it that works much better than using the previous MULTIPLE command:

1. Continue using the `12-1_Using OBJECT Commands.dwg` file.

2. Using the **Home** ribbon and the **Modify** panel, select the **Offset** command.

3. Press *Enter* to accept the **Through** command option or *left-click* to select the **Through** command option in the Command Line.

4. *Left-click* to select one of the previously drawn rectangles, and select the **Multiple** command option in the Command Line to turn this command into a repeating command.

5. *Left-click* outside the existing RECTANGLE object to offset, and continue to *left-click* as many times as needed to offset the RECTANGLE object multiple times.

   As you can see, using the **Offset | Multiple** command option involves significantly fewer clicks than the previous **Multiple** command.

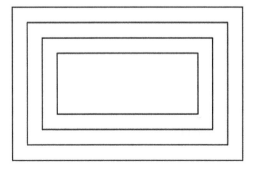

Figure 12.10: Using OFFSET Multiple

There is another very handy command option buried in the OFFSET command that is easily overlooked.

This option allows you to offset objects to the CURRENT layer instead of the same layer as the original object.

OFFSET to CURRENT Layer

In this exercise, we will offset an existing object to the current layer, not the layer of the existing object:

1. Continue using the 12-1_Using OBJECT Commands.dwg file.
2. Using the QAT toolbox, change the current layer to the **5-Dashed** layer.
3. Using the **Home** ribbon and the **Modify** panel, select the **Offset** command  and key in L for the **Layer** command option.
4. Key in C for **CURRENT** and press *Enter* to accept the **Through** command option.

Figure 12.11: OFFSET to the Current Layer

5. Select a rectangle object to offset and *left-click* to define the direction either inside or outside.

> **Note**
>
> Use the following command sequence on a button for more efficient use of these command options:
>
> ^C^C_offset;l;c;;

The next exercise will discuss how to use the FILLET command to repeat applying rounded corners to more than one corner at a time. The FILLET command has a MULTIPLE command option built in that works similarly to the previously mentioned MULTIPLE command, but it takes fewer steps.

Multiple FILLET

Another favorite of mine is the MULTIPLE option found in the FILLET command:

1. Continue the `12-1_Using OBJECT Commands.dwg` file.

2. Using the In-Canvas View Controls, restore the **Custom Model Views | 5-MULTIPLE Commands** named view.

3. Using the **Home** ribbon and the **Modify** panel, select the **Fillet** command.

4. Key in M for the **Multiple** command option, and key in R to define the fillet **Radius** as 1.

    You can also use the *right-click* menu to access the **Fillet** command options.

5. Select two of the MAGENTA corner LINE objects to apply the FILLET. Note that the command does not stop, and you can continue to round off all four corners.

    Do you know how to quickly use the FILLET command on POLYLINE objects?

6. Select the **Fillet** command again and key in P to apply the radius to all corners of the POLYLINE object automatically.

7. Select the RECTANGLE object to round off all corners.

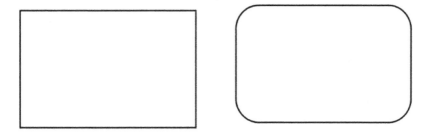

Figure 12.12: Auto-FILLET POLYLINE objects

So, how would you SQUARE OFF these same corners on the POLYLINE object?

8. Select the **Fillet** command again, and the previous RADIUS of 1 will still be applied. You do not have to reset the RADIUS to 0.

Did you know you can use the *Shift* key to temporarily set the RADIUS to 0 while running the FILLET command? Yes, but unfortunately, you can't do it with the POLYLINE command option.

1. Select the RECTANGLE object as the object we want to square off all corners.

2. Select each corner individually while using the *Shift* key to set the RADIUS to 0.

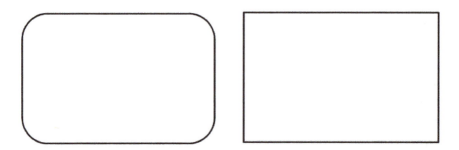

Figure 12.13: Automate the RADIUS of ZERO

> **Note**
>
> If you use the RADIUS command option to change the RADIUS to 0, the POLYLINE command option works just fine.

Here's another tip for using the FILLET command. Did you know you can FILLET CIRCLE objects or parallel lines?

1. Continue the `12-1_Using OBJECT Commands.dwg` file.

2. Using the In-Canvas View Controls, restore the **Custom Model Views | 5-MULTIPLE Commands** named view.

3. Using the **Home** ribbon and the **Modify** panel, select the **Fillet** command and *right-click* to access the RADIUS command option to set the RADIUS to `10`.

4. Select both the BLUE CIRCLE objects twice to FILLET both sides of the circles.

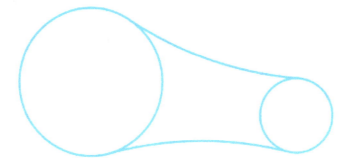

Figure 12.14: The CIRCLE fillet results

Next, we will FILLET the parallel lines:

1. Select the **Fillet** command and select two of the PARALLEL LINE objects. An ARC is added to the end of the lines closest to your selection point.

2. Continue to FILLET the parallel lines to get the following results.

Figure 12.15: The Parallel lines results

The next exercise will look at how to use the ARRAY command's hidden command options to improve productivity.

Arc Direction Change

When drawing an ARC object, there are times when it would be helpful to change the arc swing direction, right? Well, this is possible if you use the *Ctrl* key to toggle the arc direction:

1. Continue using the `12-1_Using OBJECT Commands.dwg` file.

2. Using the In-Canvas View Controls, restore the **Custom Model Views | 6-Arc Direction** named view.

3. Using the **Home** ribbon and the **Draw** panel, select the **Arc** *drop-down list* and select the **Start, Center, End** command.

4. *Left-click* to define the START of the ARC at the TOP of the LEFT LINE, and define the CENTER of the ARC at the TOP of the MIDDLE CENTER LINE.

5. *Hover* over the TOP of the RIGHT LINE, and then the ARC object will swing counter-clockwise by default. However, using the *Ctrl* key, you can draw the ARC object in the clockwise direction. *Left-click* when the ARC object swings in the desired direction.

In the next section, we will examine some of the hidden commands in the Express Tools delivered with AutoCAD.

Using the Hidden Commands

Not all hidden commands are included in the Express Tools, but there are some that I find useful and that many users overlook.

In this exercise, we will learn how to HATCH with IMAGES rather than the normal pattern objects.

SuperHatch

Every user has used the HATCH command, but have you used SUPERHATCH? SUPERHATCH is delivered with the Express Tools, which are oldies but goodies, especially if you need a quick elevation or site plan with images:

SUPERHATCH	Command Locations
Ribbon	Express Tools | Draw | SuperHatch
Command Line	SUPERHATCH (SUP)

1. Continue using the `12-1_Using OBJECT Commands.dwg` file.

2. Using the In-Canvas View Controls, restore the **Custom Model Views | 7-Super Hatch** named view.

3. Using the **Express Tools** ribbon and the **Draw** panel, select the **SuperHatch** command.

4. Using the **SuperHatch** dialog, click the **Image** button and select the following image file. Click **Open** to select the image:

   ...\ACAD_TipsTechniques\Exercise_Files\CompanyXYZ\hatch\images\ Textured_Concrete.jpg

5. Using the **Attach Image** dialog, verify the following settings:

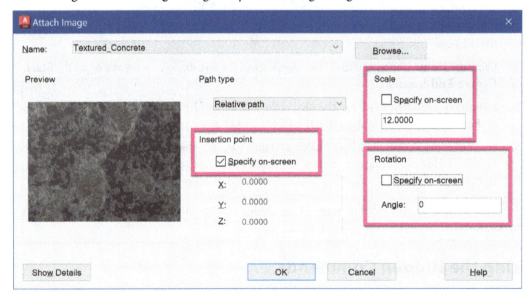

Figure 12.16: Hatch using Textured_Concrete image

6. Click **OK** to close the dialog.

7. Snap to the ENDPOINT at the LOWER-LEFT corner of the wall elevation, and select YES when asked whether the placement of the image is acceptable.

8. *Left-click* inside the wall graphics (not inside a window) to define the image hatch boundary and use the *Enter* key to complete the command.

Figure 12.17: The Wall SuperHatch

9. Select the **SuperHatch** command again, click the **Image** button, and select the following image file. Click **Open** to select the image.

    ```
    ...\ACAD_TipsTechniques\Exercise_Files\CompanyXYZ\hatch\images\
    Stained_Glass.jpg
    ```

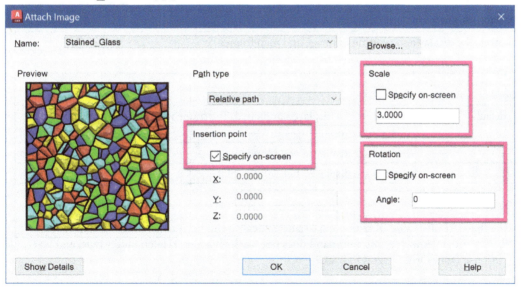

Figure 12.18: Hatch using Stained_Glass image

10. Click **OK** to close the dialog.

11. ENDPOINT snap to the LOWER-LEFT corner of the window elevation and select YES when asked if the placement of the image is acceptable.

12. *Left-click* inside all three windows to define the hatch boundaries and place the image hatch.

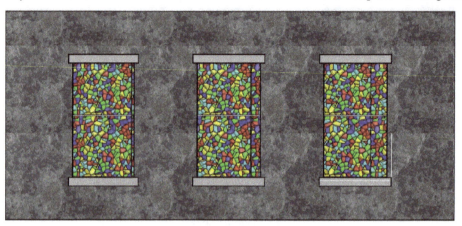

Figure 12.19: A Stained Glass SuperHatch

You can see the image frames when using SuperHatch, but we can turn their display off using the IMAGEFRAME system variable.

13. Using the Command Line, key in the IMAGEFRAME system variable and change the setting to 0.

IMAGEFRAME	
Controls the display and plotting of image and map frames.	
Type: Integer	
Saved in: Drawing	
0 (default)	Frames are not displayed or plotted. The frame will temporarily display during image selection.
1	Frames are displayed and plotted.
2	Frames are displayed but will not plot.

Note

Use the HATCHTOBACK command to automatically send normal hatches to the back of the display order. However, this command does not work on a SuperHatch object type, and you must use Draw Order | Send to Back.

Next, we want to add a break line to the HANDICAP PARKING DETAIL.

Drawing a Breakline

AutoCAD provides a BREAKLINE command to simplify drawing this common drafting annotation object. Unfortunately, it does not work with annotation scale, but hopefully, it will someday!

BREAKLINE	Command Locations
Ribbon	Express Tools | Draw | Breakline
Command Line	BREAKLINE (BREAKL)

1. Continue using the 12-1_Using OBJECT Commands.dwg file.
2. Using the In-Canvas View Controls, restore the **Custom Model Views | 8-Breakline** named view.
3. Using the **Express Tools** ribbon and the **Draw** panel, select the **Break Line Symbol** command.
4. Using the Command Line, select the **Size** command option and key in the scale of 2.
5. *Hover* over the ENDPOINT of the TOP RIGHT horizontal, and using **Polar Tracking**, slide the cursor UP and *left-click* to start the BREAK LINE object.
6. Drag the cursor DOWN and *left-click* to complete the LINE object.
7. Press *Enter* to accept the MIDPOINT break symbol location, and press *Enter* again to complete the command.

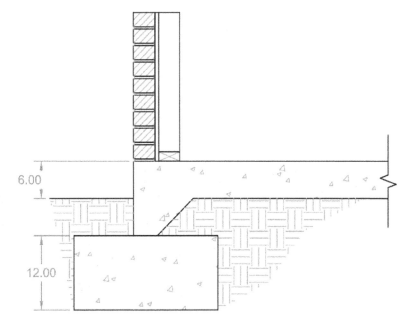

Figure 12.20: BREAKLINE symbol results

Hopefully, you know a few more Express Tools that you can put to good use.

In the next section, we will look at how to use some of the more obscure OSNAPS and their capabilities.

More object snaps (OSNAPS)

In this section, we will learn how to use some of the more obscure OSNAP modes available in AutoCAD. These OSNAPS are only available using a *right-click* menu or by keying in a shortcut in the Command Line.

In the first exercise, we will look at how to use the FROM osnap.

From

The FROM osnap allows you to place a geometry offset from a specific point in the drawing file.

In this example, we will place a circle offset from the corner of a RECTANGLE object:

1. Open the `12-2_More OBJECT Snaps.dwg` file.
2. Using the In-Canvas View Controls, restore the **Custom Model Views | 1-FROM osnap** named view.
3. Using the **Home** ribbon and the **Draw** panel, select the **Circle** command.
4. Use *Shift + right-click* to access the FROM osnap mode.
5. *Left-click* on the UPPER-LEFT corner of the RECTANGLE object and key in @-1,-1.
6. Key in the radius of .5 to complete the CIRCLE placement.

Temporary Tracking

This is easy, except that it requires us to use that dreadful @ symbol. Have you tried to do this with TEMPORARY tracking? Check this out!

1. Using the Status Bar, verify that **Polar Tracking** is enabled, and also verify that the EXTENSION osnap is turned ON.
2. Select the **Circle** command and *hover* over the UPPER-RIGHT corner of the RECTANGLE object.
3. Drag the cursor straight to the RIGHT, and you should see the "green-dashed" tracking line.
4. Key in TT for a TEMPORARY tracking point, and key in the distance of 1.

> **Note**
> You can also select TEMPORARY track point from the *right-click* menu.

5. Drag the cursor straight DOWN, and you should see the "green-dashed" tracking line and key in the distance of 1.

6. *Left-click* to accept this location for the center of the CIRCLE object, and key in the radius of .5 to complete the CIRCLE placement.

   If you find that easier, you should investigate more opportunities to use temporary tracking when placing your objects.

   Use the **Options** dialog to perfect your use of object tracking.

7. Using the Status Bar, *hover* over the **Object Snap** button 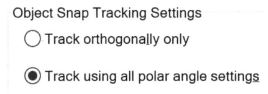 and *right-click* to access the **Object Snap Settings** command.

8. Select the **Polar Tracking** tab and review the **Object Snap Tracking Settings** options.

9. Change the setting to **Track using all polar angle settings**.

Object Snap Tracking Settings

◯ Track orthogonally only

◉ Track using all polar angle settings

Figure 12.21: Object Snap Tracking Settings

Next, we want to add smaller circles to the geometry using the MIDPOINT BETWEEN 2 POINTS osnap.

Midpoint Between 2 Points

We often need to find the "middle" of something that does not have a CENTER osnap. You can always find the middle of anything using the MIDPOINT BETWEEN 2 POINTS (M2P) osnap.

This OSNAP is overlooked by many users and should be one of your favorites:

1. Continue using the 12-2_More OBJECT Snaps.dwg file.

2. Using the In-Canvas View Controls, restore the **Custom Model Views | 2-Midpoint Between 2 Points** named view.

3. Select the **Circle** command and *Shift + right-click* to access the MIDPOINT BETWEEN 2 POINTS osnap.

4. *Left-click* on the LEFT-edge MIDPOINT of the RECTANGLE object and *left-click* on the LEFT quadrant of the MIDDLE CIRCLE object. Use *Shift + right-click* to access the QUADRANT osnap if it is not available.

5. Key in the radius of .25 to complete the CIRCLE placement.

6. Repeat *steps 1–3* on the other side of the MIDDLE CIRCLE object.

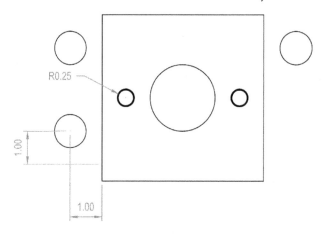

Figure 12.22: Place CIRCLES using Midpoint Between 2 Points

Next, using POLAR TRACKING, we will place another small circle below the previous two circles:

1. Select the **Circle** command, *hover* over the CENTER of the LEFT small circle, and then *hover* over the CENTER of the RIGHT small circle.

2. Moving DOWN at a **45-degree** polar angle, locate the intersecting object tracking lines to place the third CIRCLE object.

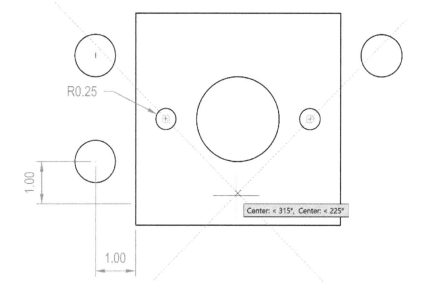

Figure 12.23: Multiple Object Tracking Lines

3. *Left-click* to accept the circle location and key in the radius of .25 to complete the CIRCLE placement.

Yes, there were other ways to draw the same circles – we could have used a ROTATE + COPY around the circle – but now you know how to use the OBJECT and POLAR TRACKING in another example.

In the next example, we will draw a centerline in the road:

1. Continue using the 12-2_More OBJECT Snaps.dwg file.

2. Using the In-Canvas View Controls, restore the **Custom Model Views | 3-Midpoint Between 2 Points** named view.

3. Using the **Home** ribbon and the **Draw** panel, select the **Polyline** command.

4. Before issuing any *left-click* points for the line, *right-click* to access the **Midpoint Between 2 Points** osnap.

5. *Left-click* on the ENDPOINT at the TOP of the FIRST road line and *left-click* again on the TOP of the SECOND road line.

6. Draw the POLYLINE object to the MIDPOINT of the corner ENDPOINTS and use the *right-click* to access the **Midpoint Between 2 Points** osnap.

7. Continue drawing the centerline object to the last two ENDPOINTS of the road lines, only this time using the M2P shortcut key to access the **Midpoint Between 2 Points** osnap.

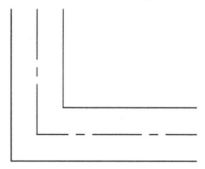

ROAD LINES

Figure 12.24: Placing LINES using Midpoint Between 2 Points results

Everyone who has ever drawn in 3D has seen those "zinger" objects that go from one elevation to another "accidentally." Let me show you how to use POINT FILTERS to eliminate this problem. I predict that very few of you are currently using POINT FILTERS, yet they are so useful! Let me help you discover the power of this tool.

In the next section, we will learn how to use the calculator provided in AutoCAD to perform simple math equations and how to use this calculator to define command values.

Using the Calculators

There are two calculators available in AutoCAD. The first is a key-in command that can be run outside or transparently inside another command. The second is the Quick Calculator (QuickCalc), which also can be run outside or inside of a command that provides a dialog calculator.

To run either calculator command inside another command, you must use the ' character as a prefix to the CAL command.

CAL	Command Locations
Command Line	CAL (CAL)

Using the CAL Command

In this example, we will use the CAL command to calculate a simple mathematical value. This is a very simple example, but if you haven't used this command before, it's a good introduction:

CAL	Command Locations
Command Line	CAL (CAL)

1. Open the `12-3_Using the Calculators.dwg` file.

2. Using the In-Canvas View Controls, restore the **Custom Model Views | 1-CAL Command** named view.

3. Using the Command Line, key in the CAL command, and then key in the `1/12` EXPRESSION to calculate the scale factor between feet and inches.

The SCALE command

Now, let's use the CAL command in the SCALE command:

1. Using the **Home** ribbon and the **Modify** panel, select the **Scale** command and then the SQUARE object.

2. Press *Enter* to complete the selection and snap to the LOWER-LEFT corner of the SQUARE object to define the base point.

3. When prompted to key in the **Scale Factor**, key in `1/12` and press *Enter* to complete the command.

> **Note**
>
> I didn't have to key in the CAL command to calculate the scale factor. You should try this "hidden" calculation feature in several commands. Any time you go to reach for your physical calculator, stop and try this instead.

In the next example, I want to calculate the offset distance using the transparent 'CAL command inside the OFFSET command.

An OFFSET example

How often have you used the DIVIDE command and placed POINT objects to calculate the offset value for additional objects? Well, stop that tedious task and use the transparent 'CAL command instead.

Don't forget to use the ' character before the CAL command to allow it to run inside the current command. Using this method, you don't even need an object to perform a divide function.

1. Continue using the `12-3_Using the Calculators.dwg` file.

2. Using the In-Canvas View Controls, restore the **Custom Model Views | 2-'CAL Command** named view.

    In this example, I placed the POINT objects just for reference to display the distance between the two vertical lines divided by 7; we don't really need them.

3. Using the **Home** ribbon and the **Modify** panel, select the **Offset** command. ⋐

4. We will use the calculator to determine the OFFSET distance.

5. Using the Command Line, key in `'CAL`, and when prompted for the EXPRESSION, key in `DEE/7` and press *Enter* to complete. Note that the offset distance has been set to `0.75`.

6. Select the LEFT vertical LINE object and *left-click* to offset it to the right.

7. Key in M for the **Multiple** command option and continue to *left-click* until all the offsets are complete, and press *Enter* to complete the command.

In this exercise, you learned how much easier this method is than using the DIVIDE command to place all those POINT objects, which you have to delete when you are finished with the OFFSET command.

So, what is **DEE/7**? It is a shortcut to calculate the distance between two ENDPOINTS and then divide that distance by 7. If you wanted to key in the long-hand version, you would key in `(dist(end,end))/7`.

There are several functions that you can use with the CAL command.

CAL Shortcut	Long-Hand Command	Description
MEE	(end+end)/2	Calculates the midpoint between two endpoints
DEE	(dist(end,end))	Calculates the distance between two endpoints
ILLE		Calculates the intersection of two points determined by four points

CAL Operators	Description
^	Indicates a numeric exponent
*	Multiplies numbers
/	Divides numbers
+	Adds numbers
-	Substracts numbers
()	Groups expressions
pi	The value of PI
round ()	Rounds off a value to the nearest integer
r2d	Converts radians to degrees
d2r	Converts degrees to radians

Quick Calculator (QuickCalc)

The Quick Calculator is a dialog version of the old physical calculators we all used in the past. This dialog provides many of the features you expect in a calculator, but it also allows interaction between objects and the calculator.

QUICKCALC	Command Locations
Ribbon	Home | Utilities | Quick Calculator View | Palettes | Quick Calculator
Right-click Menu	QuickCalc
Command Line	CALCULATOR (CALC), QUICKCALC (QC)

Let's take a look at a couple of examples.

Quick Calc from Properties

The first example will demonstrate how to use the Quick Calculator from within the PROPERTIES palette to modify existing objects. In this case, I want to modify all the CIRCLE objects so that their diameter is equivalent to the width of the ROAD objects, without using a MEASURE command:

1. Continue using the 12-3_Using the Calculators.dwg file.

2. Using the In-Canvas View Controls, restore the **Custom Model Views | 3-QuickCalc DIAMETER** named view.

3. Select one of the CIRCLE objects, *right-click* to access the **Select Similar** command to select all four CIRCLE objects and then *right-click* again to access the **Properties** command.

4. Using the **Properties** palette and the **Geometry** panel, select the **Diameter** value and then select the **Quick Calculator** icon. ▦

5. Using the **QuickCalc** dialog, click the **Distance Between 2 Points** button.

6. Using the ENDPOINT osnaps, select the TOP ENDPOINT of each vertical continuous LINE object.

7. When the **QuickCalc** dialog appears, review the distance value of 12' and click the **APPLY** button to apply this value to the diameter of both circles.

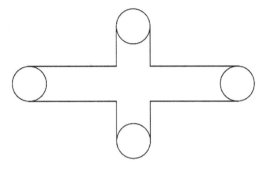

Figure 12.25: The QuickCalc Diameter results

QuickCalc Units Conversion

In this example, we will use the Quick Calculator to convert units in the design process:

1. Continue using the 12-3_Using the Calculators.dwg file.

2. Using the In-Canvas View Controls, restore the **Custom Model Views | 4-QuickCalc Units** named view.

3. Select the outside POLYLINE object and *right-click* to access the **QuickCalc** command.

4. Using the **QuickCalc** dialog, expand the **Units Conversion** panel and modify the following settings:

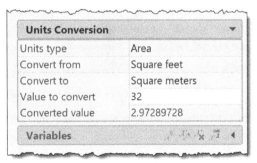

Figure 12.26: A QuickCalc Units Conversion

> **Note**
>
> Be sure to use the *Enter* key and not the *Tab* key to enter the value of 32; otherwise, the value resets to 0.

Well, I hate to do anything manually, so let's try out this next example using TEXT FIELDS to display the conversion values. I have created BLOCKS that contain TEXT FIELDS and/or ATTRIBUTES for the conversions I need most often to demonstrate this technique.

1. Using the QAT, use the **Layer** *drop-down list* and **Thaw** ☀ the **4-Objects_Fields2** layer.

    Here, you can see the TEXT FIELDS that contain the Imperial and Metric units for both AREA and perimeter LENGTH.

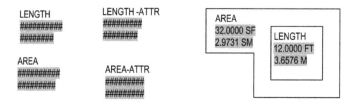

Figure 12.27: AREA and PERIMETER fields

The #### field values indicate that the TEXT FIELDS have lost their links to the original shapes, which makes sense, since these are new TEXT FIELDS that have never been linked to an object, and we want to link the BLOCK to a new object. Unfortunately, there isn't a command in AutoCAD to re-link a TEXT FIELD to a new shape without editing the FIELD – YES, there should be! Until that happens, we can use a free application that is available using the following link.

One of my favorite tools for managing fields in AutoCAD is a free application from CAD Studio: www.cadstudio.cz. This utility allows you to re-link a field to a new or existing object when the link has been lost. This is very helpful when the original object has been deleted and you have those dreaded #### field results.

> **Note**
>
> While I have successfully used this application in several versions of AutoCAD, I am not responsible for any problems that may occur with your installation.

2. Using the **Manage** tab and the **Applications** panel, select the **Load Application** command. 🔧

3. Using the **Load/Unload Applications** dialog, navigate to the following location and select the **ReLinkField** application:

    ...\ACAD_TipsTechniques\Exercise_Files\CompanyXYZ\support\

4. Click the **Load** button, and then click on **Close** to close the dialog.

5. Using the **Insert** ribbon and the **Block** panel, select the **Insert | AREA Unit Conversion** block and place the block inside the empty POLYLINE shape. This block must be exploded for the new **ReLinkField** command to work.

6. Using the **Home** Ribbon and the **Modify** panel, select the **Explode** command 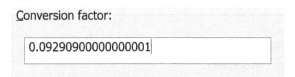 and select the newly placed block.

7. Using the Command Line, key in the new RLF command and select the block that contains the missing links. Then, select the new SHAPE object to link to. The TEXT FIELDS should update automatically.

Next, we will insert a block that uses ATTRIBUTES that contain the same TEXT FIELDS. This block type does not need to be exploded to be re-linked to an object:

1. Using the **Insert** ribbon and the **Block** panel, select the **Insert | AREA Unit Conversion-ATTR** block and place it inside the empty POLYLINE shape.

2. Using the Command Line, key in the new RLF command and select the block that contains the missing links. Then, select the new SHAPE object to link to. The TEXT FIELDS again update automatically.

Both of these blocks provide a quick method for conversion values, especially if you need them as text in a drawing file.

Note

The TEXT FIELDS in these blocks were created using **Edit Field | Object | Area | Additional Format | Conversion Factor.**

Conversion factor:

0.09290900000000001

Figure 12.28: A Text Field Conversion Factor

As a general rule, TEXT FIELDS do not update automatically. By default, they update based on the FIELDEVAL system variable.

FIELDEVAL	
Controls when FIELDS update.	
Type: Integer	
Saved in: Drawing	
Initial Value: 31	
0	Fields are not updated.
1	Fields are updated when a file is opened.
2	Fields are updated when a file is saved.
4	Fields are updated when a file is plotted.
8	Fields are updated when you use the ETRANSMIT command.
16	Fields are updated when you use the REGEN command.
Note: The variable is stored as a bitcode that uses the sum of these values.	

This is a good time to discuss how this type of system variable works. This system variable is a BITCODE variable, which is defined by adding the available values together to determine which settings are enabled.

For example, the default setting is 31 (1+2+4+8+16), which is the sum of all the available values:

> **Note**
>
> If you only want FIELDS to update when you open a drawing, you would set this system variable to 1.
>
> If you want FIELDS to update when you open or plot a file, you would set this system variable to 5 (1+4).

With these examples, you should be able to apply the Quick Calculator utility to many areas in your own projects and get creative. Good luck!

In the next exercise, I want to show you how you can "tweak" how the SELECT SIMILAR command works.

More Bonus Commands

Using Select Similar

Hopefully, everyone uses the SELECT SIMILAR command, but does it always work the way we want? How many times have you wanted to select objects using a specific DIMENSION STYLE but wanted to ignore the layer setting? Yes, we can do that!

Most users do not know about this command option because we always use SELECT SIMILAR from the *right-click* menu, rather than from a command key, so the SETTINGS option is hidden:

1. Open the 12-4_Using BONUS Commands.dwg file.

2. Using the In-Canvas View Controls, restore the **Custom Model Views | 1-Select Similar** named view.

3. Select one of the exterior dimensions and *right-click* to access the **Select Similar** command.

    As expected, only the exterior dimensions are selected because the interior dimensions are on a different layer.

4. Using the Command Line, key in the SELECTSIMILAR command and select the **Settings** command option.

5. Turn OFF the **Layer** setting, leaving the **Name** setting ON so that only the Dimension Style name is used for the selection. Click **OK** to close the dialog.

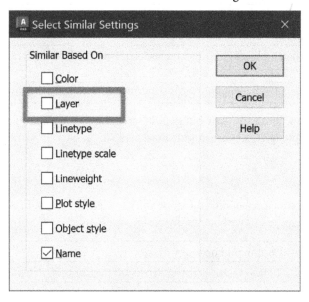

Figure 12.29: Select Similar Settings

6. Select one of the exterior dimensions and *right-click* to access the **Select Similar** command.

Now, all dimensions using **Dimension Style | Standard** are selected because the layer is ignored. Use this option when you want to quickly make global changes to all the objects using a specific property in a drawing file.

Use the SELECTSIMILARMODE system variable to define your preferred Select Similar settings.

SELECTSIMILARMODE	
Controls which properties to match when using the SELECTSIMILAR command.	
Type: Bitcode	
Saved in: User-Settings	
Initial Value: 130	
0	Match Object TYPE
1	Match Object COLOR
2	Match object LAYER
4	Match object LINETYPE
8	Match object LINETYPE SCALE
16	Match object LINEWEIGHT
32	Match object PLOT STYLE
64	Match object STYLE, such as TEXT STYLES, DIMENSION STYLES, and TABLE STYLES
128	Match object name, such as REFERENCES, BLOCKS, and IMAGES

In the next exercise, we will use an "oldie but goodie" command to UNDO just the last ERASE or DELETE command, skipping all other manipulations.

OOPS

We all use the UNDO command, but did you know you can UNDO the last ERASE using the OOPS command, and it will ignore all other commands except for ERASE?

OOPS	Command Locations
Command Line	OOPS (OO)

1. Continue using the `12-4_Using BONUS Commands.dwg` file.
2. Using the In-Canvas View Controls, restore the **Custom Model Views | 2-OOPS** named view.
3. Using the **Home** ribbon and the **Draw** panel, select the **Offset** command and offset the RECTANGLE object to the inside twice, using a distance of `.25`.
4. Select the **ERASE** command ✎ to remove the MIDDLE OFFSET object.
5. Select the **Line** command and draw an **X** inside the smallest RECTANGLE object.

If I want to restore the MIDDLE OFFSET rectangle object, I can't use the UNDO command, as it will undo the two crossing LINES. If I use the OOPS command, I can restore just the MIDDLE OFFSET rectangle object.

6.   Using the Command Line, key in OOPS and press *Enter* to complete the command.

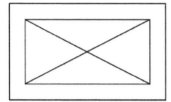 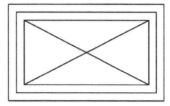

Figure 12.30: The OOPS Command

In the next exercise, we will use the NUDGE command to slightly move an object.

NUDGE

The NUDGE command allows you to move objects in very small increments without using the MOVE command. To access the NUDGE command, first, select the object you want to move and use the *Ctrl + Arrow* keys to move the object in any of the four arrow directions. This movement is orthogonal only:

1.   Continue using the 12-4_Using BONUS Commands.dwg file.

2.   Using the In-Canvas View Controls, restore the **Custom Model Views | 3-NUDGE** named view.

3.   Select the first SQUARE object, and use the *Ctrl + Arrow Right* keys to "nudge" the SQUARE object to the RIGHT by a distance of .1.

4.   Continue to use the *Ctrl + Arrow Right* keys to NUDGE the objects as needed.

5.   If you want to control the NUDGE distance, use the **Snap Spacing**.

6.   Using the Status Bar, *right-click* on the **Osnap** button to access **Object Snap Settings**.

7.   Using the **Drafting Settings** dialog, select the **Snap and Grid** tab, locate the **Snap Spacing** section, and turn ON the **Snap ON** setting.

8.   Also, turn ON the **Equal X and Y spacing** setting and set the **X** value to .25. Then, click the **Close** button to close the dialog.

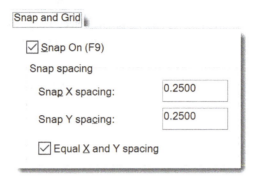

Figure 12.31: Object Snap Settings

9. Select the last SQUARE object, and use the *Ctrl + Arrow Left* keys on the keyboard to "NUDGE" the SQUARE to the left.

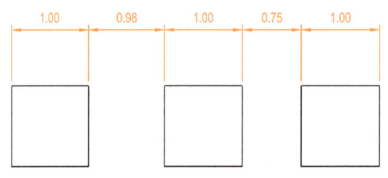

Figure 12.32: Using NUDGE results

The Reverse Command

Use the REVERSE command to change the direction an object was drawn in. We no longer need this for linetypes since the introduction of the automatic READ-WRITE linetype direction option, but you might need this for linetypes that use symbols, such as arrows:

REVERSE	Command Locations
Ribbon	Home | Modify | Reverse
Command Line	REVERSE (REVE)

1. Continue using the `12-4_Using BONUS Commands.dwg` file.

2. Using the In-Canvas View Controls, restore the **Custom Model Views | 4-Reverse Command** named view.

3. Using the **Home** ribbon and the **Modify** expanded panel to locate the **Reverse** command. ⇄

4. Select the TOP LINE object and press *Enter* to complete the command.

    The line is now drawn in the opposite direction.

5. Repeat *steps 3–4* for the remaining two LINE objects.

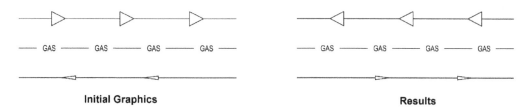

Initial Graphics   **Results**

Figure 12.33: REVERSE command

You no longer need to worry about custom linetypes with text, as they are now controlled with a new setting in the linetype definition file (.LIN).

Note

The REVERSE command option has been added to the PEDIT command.

Rotating by Direction

Everyone uses the ROTATE command and is familiar with rotating objects using the standard 0-360-degree angles. Did you know you can also use directions?

Instead of using 90, 180, and so on, you can key in N, W, S, or E to rotate objects based on a compass direction:

-   N = 90 degrees

-   S = 270 degrees

-   E = 360 degrees

-   W = 180 degrees

These directions can be affected by your UNITS settings if you have changed the default angle measurements from counter-clockwise to clockwise, but that only changes the directions to the reversed angles.

Note

If you want to rotate an object -270 degrees, you can key in ‑S.

Let's try this out. Here, four arrow shapes point in the E direction. We will rotate them one at a time to fill in our compass:

1. Continue using the `12-4_Using BONUS Commands.dwg` file.

2. Using the In-Canvas View Controls, restore the **Custom Model Views | 5-Rotate Like Map** named view.

3. Using the **Home** ribbon and the **Modify** panel, select the **Rotate** command. ↻

4. Select the ARROW shape and use *Enter* to complete the selection.

5. Snap to the NODE at the center of the CIRCLE as the base point and key in N for **North** to rotate the shape 90 degrees.

6. Select the next original ARROW shape and press *Enter* to complete the selection.

7. Snap to the NODE at the center of the CIRCLE as the base point and key in W for **West** to rotate the shape 180 degrees.

8. Select the next original ARROW shape and press *Enter* to complete the selection.

9. Snap to the NODE at the center of the CIRCLE as the base point and key in S for **South** to rotate the shape 180 degrees.

I'm not sure if this saves you time or not, but I was surprised to discover this aspect of the ROTATE command.

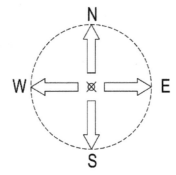

Figure 12.34: Rotate by Direction results

Open to Named View

We have used NAMED VIEWS throughout this course, but do you use them in your projects? You can do so much with named views to control your project displays that I want to show you another reason to use them.

Did you know that you can open a file to a specific named view? Check this out!

1. Using the **QAT** toolbar, select the **Open** command.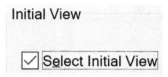

2. Using the **Select File** dialog, select the 12-5_Opening Named Views.dwg file, locate the **Initial View** section, and turn on **Select Initial View**.

Initial View

☑ Select Initial View

Figure 12.35: Opening the Initial View (the Named View)

3. Click the **Open** button to display the Named Views stored in this drawing file.

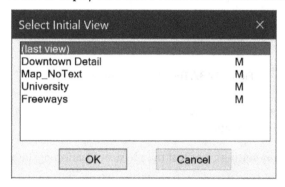

Figure 12.36: Opening Using a Named View

4. Using the **Select Initial View** dialog, note that the default view is the last view displayed when the file was saved, but we can select any Named View.

5.   Select the **University** Named View and click **OK** to open the file using this view.

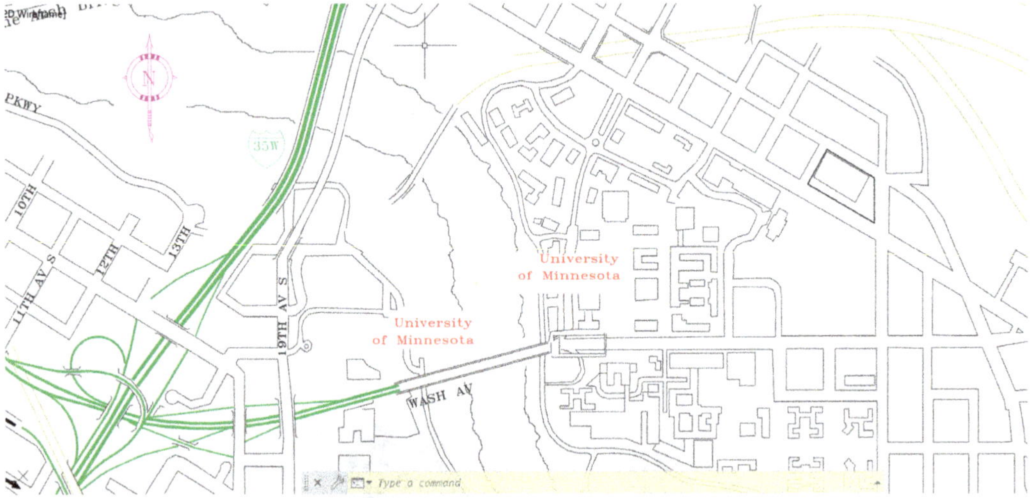

Figure 12.37: The University Named View

6.   Repeat the **Open** command and select the **Freeways** Named View. Note that the **Initial View** setting is turned OFF automatically after each use.

A Named View stores many settings, not just the physical location of the view. It also stores the following view settings:

- Paperspace/Modelspace
- Physical Location
- User Coordinate System (UCS)
- Layers On/Off
- View Category (Sheet Sets)
- Perspective
- Live Section
- Visual Style
- Background
- Annotation Scale
- Clipping Planes

Now that you know how to use a Named View when opening a file, you can try using it in your projects.

Summary

In this chapter, we learned how to use commonly used commands beyond our everyday usage. We explored TRIM, PEDIT, and ARRAY commands in more detail. We learned how to turn these common commands into repeating commands for more productivity. We learned how to take advantage of the built-in calculators inside AutoCAD.

In the next chapter, we will examine how to use drawing file utilities to improve performance and fix corrupt or oversized files, as well as understand system variables to improve our AutoCAD system for personal use.

13

Using Commands
to Improve Performance

This chapter will teach you how to use many utility commands to improve your drawing performance by cleaning up unused and duplicated data and how to repair a file using the AUDIT and RECOVER commands. You will also learn how to use system variables and how they can personalize your AutoCAD experience. We will also learn how to use some of the hidden features of viewports, such as Merge, Rotate, and Change Space. You will learn how to automate and control those custom linetypes and how to use more Bonus Tools.

In this chapter, we'll cover the following topics:

- Drawing performance
- Repairing files
- Using linetype scales
- Advanced viewport options
- Bonus commands

By the end of this chapter, you will be able to maintain efficient and clean drawing files, work more efficiently with viewports, and customize and maintain AutoCAD using system variables.

Drawing utilities

In this section, we will investigate how you can maintain your drawing files to keep them working efficiently. These utilities include removing unused data to keep your files as small and clean as possible. We will learn how to remove unused application data and duplicated and/or sloppy graphics.

Drawing performance

Improving the performance of your drawing files is a simple but necessary task everyone should perform on a regular basis. This maintenance doesn't happen automatically. Every user needs to perform these tasks regularly to keep their files working at their optimal efficiency.

PURGE REGAPPS

Today, it is very common to receive files from other Autodesk applications, such as Civil 3D, Architectural Desktop, Inventor, and so on. These files often include specific objects created using these applications. If you use AutoCAD and do not need these intelligent objects, you can purge these application objects from the drawing file to improve performance. However, this option does not appear in the PURGE dialog.

PURGE	Command Locations
Application Menu	"Application A" | Drawing Utilities | Purge
Ribbon	Home | Modify | Delete Duplicate Objects
	Manage | Cleanup | Purge
Menu	File | Drawing Utilities | Purge
Command Line	PURGE, -PURGE (PU)

Use the -PURGE command, the key-in version of the PURGE command, to purge the REGAPPS from your files. REGAPPS are objects that are created using other Autodesk applications.

1. Open the 13-1_CLEANUP Commands.dwg file.

2. Using the In-Canvas View Controls, restore **Custom Model Views | 1-REGAPPS** named view.

3. Using the Command Line, key in the -PURGE command and select the **REGAPPS** command option.

4. Press the *Enter* key to accept the <*> wildcard selection for all REGAPPS, and key in N for *NO* to verify each regapp being removed.

> **Note**
>
> Using the - character for any command runs the legacy or command-line version of the command to avoid any dialog interactions. This is very useful when creating custom interface items using custom command strings.

The PURGE command removes one level of nested items at a time. Repeat the command to remove additional nested items. I run the -PURGE command three times to remove all nested items; however, depending on your organization's nested items, you might need more repetitions.

Non-purgeable objects

Using the PURGE command dialog, you often have non-purgeable items that you can't get rid of. Finding these objects can be challenging. You think they are all gone, but there are one or two left in the drawing that you can't find. Have you tried using the new SELECT OBJECTS button to locate these items?

1. Continue using the 13-1_CLEANUP Commands.dwg file.

2. Using the In-Canvas View Controls, restore **Custom Model Views | 2-Non-Purgeable Items** named view.

3. Using the **Manage** ribbon and the **Cleanup** panel, select either the **Find Non-Purgeable Items** command or the **Purge** command. 🔄

4. Using the **Named Item** list, expand the **Blocks** section and select the COMPUTER block.

5. Use the **Details** panel on the RIGHT side of the dialog to review where the block is used. Here, you can see that there are still two blocks remaining in the file.

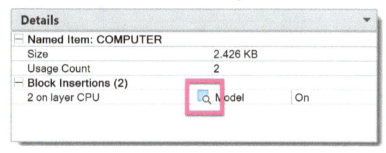

Figure 13.1: Select Non-Purgeable Items

6. Click on the SELECT OBJECTS button 🔍 and the view is zoomed to locate the remaining blocks.

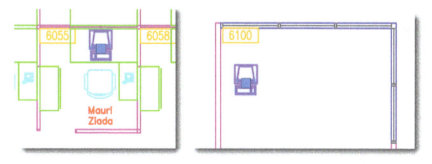

Figure 13.2: Non-purgeable blocks location

In the next exercise, we will use the OVERKILL command to clean up a drawing file.

OVERKILL

Use the OVERKILL command to remove duplicate and/or overlapping objects, such as lines, arcs, and polylines. This command will combine partially overlapping or contiguous objects where possible.

The following changes are made to the objects in your drawing file:

- Duplicate objects are deleted
- Overlapping arcs on circles are deleted
- Partially overlapping lines are combined into a single line
- Overlapping and zero-length linear segments are deleted

OVERKILL	Command Locations
Ribbon	Manage | Cleanup | Overkill
	Home | Modify | Delete Duplicate Objects
Menu	Modify | Delete Duplicate Objects
Command Line	OVERKILL, -OVERKILL (OV)

In this example, we will clean up the duplicate and overlapping objects:

1. Continue using the 13-1_CLEANUP Commands.dwg file.

2. Using the In-Canvas View Controls, restore **Custom Model Views | 3-Overkill** named view.

    Now, you can see that there are several duplicate and overlapping objects. I intentionally assigned various colors and layers to the overlapping objects to clarify them and give them a more visual display.

3. Using the status bar, turn ON SELECTION CYCLING (*Ctrl + W*) to gain visual clues when *hovering* over the overlapping and duplicate objects.

    Make note of the cursor badge that appears near the cursor when overlapping or duplicate objects are detected using selection cycling.

> **Note**
>
> Use the Temporary Override shortcut, *Shift + Spacebar*, to activate selection cycling without the status bar.

4. Using the **Manage** ribbon and the **Cleanup** panel, select the **Overkill** command.

5. *Hover* over the COMPUTER block to see that several blocks are on top of each other. Select all the COMPUTER blocks and use the *Enter* key to complete the selection.

6. Using the Command Line, we see three duplicate blocks.

7. Using the **Overkill** dialog, click the **OK** button to delete the duplicate blocks.

8. Again, using the Command Line, you can see that only one duplicate block was deleted.

```
Command: _overkill
Select objects: Specify opposite corner: 3 found
Select objects:
1 duplicate(s) deleted
0 overlapping object(s) or segment(s) deleted
```

Figure 13.3: Overkill duplicate blocks results 1

Why was only one of the duplicates removed? Because the blocks are not all on the same layer.

1. Use *Ctrl + Z* to **UNDO** the previous OVERKILL command.

2. Select the **Overkill** command again, and select all the **COMPUTER** blocks.

3. Using the Command Line, we see three duplicate blocks.

4. Using the **Overkill** dialog and the **Ignore Object Property** section, turn the **Layer** setting on so the different layers are ignored.

5. Click the **OK** button to delete the duplicate blocks.

6. Using the Command Line, you can see that both duplicate blocks are deleted.

```
Command: _overkill
Select objects: Specify opposite corner: 3 found
Select objects:
2 duplicate(s) deleted
0 overlapping object(s) or segment(s) deleted
```

Figure 13.4: OVERKILL duplicate blocks results 2

Next, we want to clean up the overlapping LINE objects. First, let's review the LINE objects involved:

1. Continue using the 13-1_CLEANUP Commands.dwg file.

2. *Hover* over the horizontal LINE objects at the POINT location and, using the **Selection Cycling** dialog, review the four LINE objects. The two BLACK and GREEN lines are on the 3-Objects layer, and the ORANGE line is on the 3-Object layer. The BLACK line objects are co-linear, connecting from end to end. The GREEN and ORANGE lines overlap.

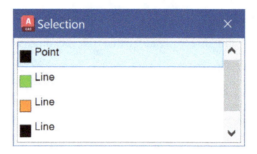

Figure 13.5: OVERKILL overlapping objects selection

3. Select the **Overkill** command and select all the horizontal LINE objects.

4. Using the **Overkill** dialog and the **Ignore Object Property** section, turn the **Color** and **Layer** settings on so that both properties are ignored.

5. Click the **OK** button to delete the overlapping objects and join all four lines into a single LINE object.

6. Select the remaining LINE object and review its properties. The resulting LINE object adopts the properties of the **DRAW ORDER | FRONT** object, but this is not consistent in all situations.

```
Command: _overkill
Select objects: Specify opposite corner: 4 found
Select objects:
0 duplicate(s) deleted
3 overlapping object(s) or segment(s) deleted
```

Figure 13.6: OVERKILL overlapping objects results 1

Another great feature of the OVERKILL command is the ability to reduce the number of vertices in a polyline, creating a much simpler POLYLINE object:

1. Continue using the 13-1_CLEANUP Commands.dwg file.

2. Select the multi-vertex POLYLINE object to review how many vertices are in this POLYLINE.

Figure 13.7: Initial multi-vertex POLYLINE

3. Select the **Overkill** command and multi-vertex POLYLINE object.

4. Using the **Overkill** dialog and the **Ignore Object Property** section, turn the **COLOR** and **LAYER** settings off so that both properties are evaluated.

5. Click the **OK** button to eliminate all the vertices in the POLYLINE object.

6. Select the new POLYLINE object to review the new simplified vertices.

Figure 13.8: Multi-vertex POLYLINE results

> **Note**
>
> OVERKILL and the DELETE DUPLICATE OBJECTS are the same command: OVERKILL.

Next, we want to delete the overlapping ARC on TOP of the CHAIR block object.

Using the OVERKILL command, select the CHAIR block and the ORANGE ARC objects. Nothing is detected since they are not exact graphic duplicates or overlaps.

In the next exercise, we will learn how to analyze your AutoCAD software when it seems slower than expected or doesn't respond at the level you anticipate.

Performance Analyzer

The new Performance Analyzer helps you to understand what issues may affect your AutoCAD performance. Performance issues occur during the following:

- Starting and opening the application or drawing files
- Mouse and cursor movement is slow and non-smooth
- Switching between layouts, model space, and paper space
- Plotting, either model space or paper space
- During the operation of specific commands
- Selecting objects in a large file

In this example, we will use the analyzer to monitor performance during your project work.

1. Open the 13-2_ANALYZE Drawing.dwg file.

2. Using the **Manage** ribbon and the **Performance** panel, select the **Performance Analyzer** utility.

3. Click on the **Start Recording** button, and click **YES** to allow this application to change your device.

4. With the recording running, perform the operations that are causing you difficulty, such as the following:

    - Starting AutoCAD

    - Opening a drawing

    - Run problematic commands

    - Execute printing functions

> **Note**
>
> Use the auto-hide feature ◄ to minimize the Performance Analyzer palette.

5. When your testing is complete, click the **Stop Recording** button.

Recording using the Performance Analyzer has a 60-minute time limit, after which it will stop automatically. The Performance Analyzer uses a large amount of disk space and memory, so it is not recommended that you run it for long periods of time. A red dot will appear next to the cursor when the recorder is running.

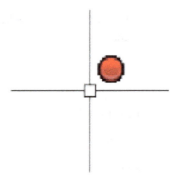

Figure 13.9: The RECORD badge icon

1. Use the **File Access Times** and **Regen Times** tabs to review the performance results. When the access times are less than .1 seconds, the results are not displayed in the palette.

> **Note**
>
> The Performance Analyzer utility requires administrator privileges.

In the next exercise, we will use the OVERKILL command to clean up a drawing file.

Repairing damaged files

Several commands can be used to repair a damaged AutoCAD file: AUDIT, REVCOVER, RECOVERALL, and RECOVERAUTO.

AUDIT

The AUDIT command is used to evaluate objects in the current space, model space, or paper space for errors and attempt to repair the drawing file. All objects found with errors are placed in a selection set, which you can access using the PREVIOUS selection command. If the AUDITCTL system variable is set to 1, a text file is generated with an (.ADT) extension that describes the problem and what action was taken. If the drawing file contains errors that the AUDIT command cannot repair, you can use the RECOVER command to retrieve the drawing and correct the errors.

AUDIT	Command Locations
Application Menu	"Red A"	Drawing Utilities	Audit
Ribbon	Manage	Cleanup	Audit
Menu	File	Drawing Utilities	Audit
Command Line	AUDIT (AU)

> **Note**
>
> The PREVIOUS command is available in any command that provides the Select Objects prompt by keying in P.

In this exercise, we will learn how to audit and clean up a drawing file:

1. Continue using the 13-2_ANALYZE Drawing.dwg file.
2. Using the **Manage** ribbon and the **Cleanup** panel, select the **Audit** command.
3. When asked, **Fix any errors detected?**, key in Y for *yes*.

RECOVER	Command Locations
The RECOVER command performs an AUDIT and attempts to open a drawing file.	
Application Menu	"Red A"	Drawing Utilities	Recover	Recover
Command Line	RECOVER (RECOVER)

RECOVERALL	Command Locations
The RECOVERALL command audits the current drawing and evaluates all reference files. The results are displayed in the Drawing Recovery palette.	
Application Menu	"Red A" | Drawing Utilities | Recover | Recover with Xrefs
Command Line	RECOVERALL (RECOVERALL)

RECOVERAUTO

The RECOVERAUTO system variable controls the display to control how notifications are displayed before and after recovering a damaged drawing file.

RECOVERAUTO	
This defines the active project name for the current drawing. This project must be defined using Options | Files | Project Files Search Path. The ProjectName points to a registry setting that includes one or more search paths for each project defined in Options.	
Type: Bitcode	
Saved in: Registry	
0 (default)	This function displays a notification to recover a damaged drawing. After the file is recovered, the drawing is opened, and a report is displayed in a dialog box.
1	This function automatically recovers the damaged drawing, opens the drawing, and displays a report in a dialog box. If a script is running, the notification is not displayed.
2	This function displays a notification to recover a damaged drawing. After the file is recovered, it opens the drawing and displays a report in the Command window.
3	This function automatically recovers the damaged drawing, opens the drawing, and displays a report in the Command window.

Using system variables

A **system variable** is a setting that controls the behavior of AutoCAD commands. They contain "default" values for their operation and can even control the appearance of the user interface.

SETVAR

To get a list of all system variables, use the SETVAR command or the new System Variables command in the Express Tools:

1. Open the 13-3_AutoCAD Configuration.dwg file.
2. Using the In-Canvas View Controls, restore **Custom Model Views | 1-HPLAYER** named view.

3. Using the **Express Tools** ribbon and the **Tools** panel, select the **System Variables** command.

4. Using the **System Variables** dialog, review the listed system variables.

5. Select the HPLAYER system variable to view the current and available settings.

> **Note**
>
> Use the FILTER field in the **System Variables** dialog to simplify finding a specific variable.

HPLAYER	
This specifies a default layer for new hatches and fills in the current drawing. You can specify a non-existent layer as the default layer for new hatches and fills.	
Type: String	
Saved in: Drawing	
. (default)	Use current layer
<layername>	Use the defined <layername> for new hatches and fills

1. Using the **System Variables** dialog, verify that the HPLAYER variable is selected and key in the layer name 1-HATCHES in the **New Value** box.

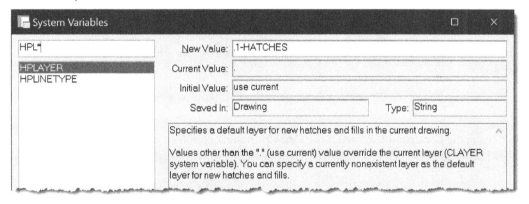

Figure 13.10: Define the HPLAYER system variable

2. Using the QAT, use the **Layer** *drop-down list* to review existing layers in this drawing file. The layer 1-HATCHES does not exist.

3. Using the **Home** ribbon and the **Draw** panel, select the **Hatch** command.

4. Select the **AR-CONC** hatch pattern and verify that the **SCALE** value is set to 1.

5. Using the ribbon, select the **Pick Points** command option and *left-click* inside the closed RECTANGLE and FLOOR POLYLINE objects to place the CONCRETE hatch pattern.

6. Use the *Enter* key to complete the command.

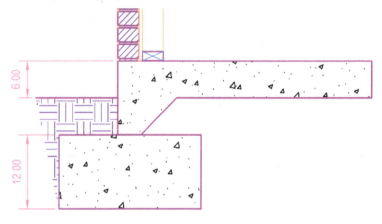

Figure 13.11: The HPLAYER system variable results 1

7. Select the new CONCRETE pattern and note that it is on the 1-HATCHES layer, created based on our HPLAYER system variable.

If you want to control more properties than just the layer name, such as line weight, color, and so on, you should create the layer, set its properties in the drawing file, and define that layer for the HPLAYER system variable.

In this example, we will use the HATCH | DRAW command option, which is often overlooked by many users:

1. Select the **Hatch** command and select the **EARTH** hatch pattern. Verify that the **SCALE** value is set to 12.

2. Using the Command Line, select the **Draw** command option and draw an area you want to fill with the EARTH hatch pattern, as shown here.

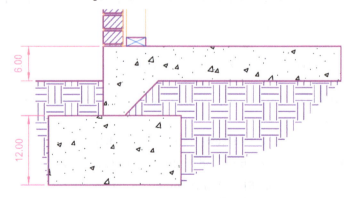

Figure 13.12: The HPLAYER system variable results 2

3. Use the *Enter* key to complete the command.

Using Wildcards	
When searching for a system variable, you can use the following "wildcard" characters to find the one you need.	
*	Matches any characters and can be used anywhere in the search string. Example: PE* matches PEDITACCEPT, PELLIPSE, and so on.
#	Matches any numeric digit found in the search string. Example: #* matches 3DCONVERSIONMODE, 3DDWFPREC, and so on.
@	Matches any alphabetic character in the search string. Example: EDGE@ODE matches EDGEMODE.
. (Period)	Matches any non-alphanumeric character in the search string. Example: .* matches *TOOLPALETTEPATH, _PKSER, and so on.
?	Matches any single character. Example: ?BC matches ABC, 3BC, and so on.
~ (Tilde)	Matches anything but the pattern, similar to exclude these characters wildcard. Example: ~*AB*matches all strings that don't contain AB.
[xx]	Matches any one of the characters enclosed in the brackets. Example: [AB]C matches AC and BC.
[~]	Matches any character not enclosed in the brackets. Example: [~AB]C matches XC but not AC.
[-]	Specifies a range for a single character. Example: [A-G]C matches AC, BC, and so on to GC, but not HC.
` (Reverse Quote)	Reads the next character literally. Example: `~AB matches ~AB.

More about system variables

There are several types of system variables, such as DRAWING, REGISTRY, USER SETTINGS, and BITCODES.

Drawing type

Drawing-type system variable values are stored in the drawing file.

Registry type

Registry-type system variable values are stored in the Windows registry and affect all drawings.

User Settings type

User Settings system variable values are stored for each individual user.

Bitcode type

Bitcode-type system variable values are defined using a unique combination of behaviors based on their accumulated values.

Let's look at a couple of system variables that are commonly used and can be inadvertently changed while using other commands.

PICKADD

The first system variable is the PICKADD system variable, which controls how you perform selections and how selection sets are maintained. This system variable can also be modified using the PICKADD button located at the top of the PROPERTIES palette. This button allows you to easily turn off the PICKADD selection, by mistake, which honestly is most often clicked by mistake.

PICKADD	
This controls how selections are created and used. Determines if subsequent selections clear an existing selection or add to it.	
Type: Integer	
Saved in: User Settings	
0	Disables PICKADD. When adding objects to an existing selection set, previously selected objects are removed from the selection set. When the *Shift* key is used to add objects to the current selection set, and no objects are selected, the previously selected objects are automatically removed from the selection set.
1	Enables PICKADD. When adding objects to an existing selection set, previously selected objects remain part of the selection set.
2 (default)	Enables PICKADD. Selected objects are added to the current selection set. If the SELECT command is used, objects remain selected after the command ends.
3	Disables PICKADD. When adding objects to an existing selection set, previously selected objects are removed from the selection set. When *Shift* is used to add objects to the current selection set, and no objects are selected, the previously selected objects remain part of the current selection set.
Note: Use the *Shift* key to add or remove objects from a selection set.	

System Variable Monitor

The System Variable Monitor contains system variables you want to monitor for unintended changes. The SYSVARMONITOR command is only available as a key-in.

When you start customizing your system variables, you may find that some commands alter your preferred settings without letting you know it is happening. Then, you move on to other commands, and things don't work how you expect them to.

For example, we have all experienced the dreadful change to the PICKADD variable we discussed previously, as well as a couple of other variables that change without notice, such as **FILEDIA** and the **PICKFIRST** variables.

FILEDIA	
This controls whether the file dialogs display when set to (1) or do not display when set to (0) for commands such as File | Open.	
Type: Integer	
Saved in: Registry	
0	Disables the display of dialog boxes for file navigation.
1 (default)	Enables the display of dialog boxes for file navigation.

PICKFIRST	
This controls whether you can use the VERB/NOUN or NOUN/VERB selection sequence, which allows you to select the command first and then select objects (VERB/NOUN) or to select objects first and then the command (NOUN/VERB).	
Type: Integer	
Saved in: Registry	
0	Turns off the ability to select objects before you select a command.
1 (default)	Turns on the ability to select objects before you select a command.

So, let's add these variables to be monitored using the System Variable Monitor:

1. Continue using the `13-3_AutoCAD Configuration.dwg` file.

2. Using the In-Canvas View Controls, restore **Custom Model Views** \| **2-EDGEMODE** named view.

3. Using the Command Line key, in the `SYSVARMONITOR` command.

You will notice that the PICKADD, FILEDIA, CMDDIA, and PICKFIRST variables are already included in the out-of-the-box monitor. So, let's add a couple of additional system variables that I like to monitor, such as EDGEMODE, TRIMEXTENDMODE, and PEDITACCEPT:

1. Using the **System Variable Monitor** dialog, verify the **Notify when these system variables change** setting and the **Enable balloon notification** settings are checked ON and click the **Edit List…** button.

2. Using the **Edit System Variable** dialog and the **Available System Variables** list, use the **Search List** filter to locate the **EDGEMODE** variable quickly.

3. Select the EDGEMODE variable and click the ADD button ⟶ >> to add this variable to the Monitored System Variable List.

4. Continue by adding the **TRIMEXTENDMODE** and **PEDITACCEPT** variables, and click **OK** to close the dialog.

5. Modify the settings for **EDGEMODE**, **TRIMEXTENDMODE**, and **PEDITACCEPT** and use them to match the following:

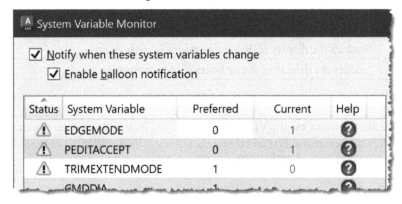

Figure 13.13: The SYSVARMONITOR settings

6. Click **OK** to close the dialog and save these changes.

7. Using the **Home** ribbon and the **Modify** panel, select the **Trim** command and select the horizontal LINE objects to trim.

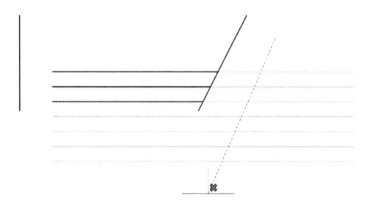

Figure 13.14: TRIM using QUICK mode results

Yes, this is easier, unless you need the TRIM command to recognize the vertical lines during the trim operation, like it did in the previous versions with EDGEMODE enabled.

Use the EDGEMODE system variable to extend linework using "imaginary" edges during TRIM and EXTEND operations.

This behavior changed in AutoCAD 2021 when the TRIM and EXTEND commands were streamlined, and the new TRIMEXTENDMODE system variable was introduced, providing a QUICK and STANDARD trim mode. Using the new streamlined TRIM and EXTEND, objects are automatically selected as cutting or boundary edges.

EDGEMODE	
This controls how the TRIM and EXTEND commands determine cutting and boundary edges.	
Type: Integer	
Saved in: Registry	
0 (default)	Allows you to TRIM or EXTEND to existing edges only without any extensions.
1	Allows you to TRIM or EXTEND to non-existing extension edges.

The downside of this enhancement is objects no longer see those "imaginary" edges. We can change this behavior using the new MODE command option to change the TRIM command back to the STANDARD method of operation:

1. Use the **UNDO** (*Ctrl + Z*) command to restore the previous graphics.

2. Select the **Trim** command, then select the **Mode** command option, and change it to **Standard**.

3. Select the horizontal LINE objects to trim and notice the difference using the STANDARD mode. However, all of the horizontal lines still do not trim past the physical line edges.

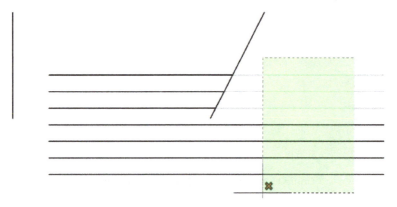

Figure 13.15: TRIM using STANDARD mode results

4. Using the Command Line, key in the EDGEMODE command and set it to 1 to enable it.

When you change this system variable, the System Variable Monitor will notify you of this change.

Figure 13.16: System Variable Monitor

5. Use the **Click to View Change** link to open the System Variable Monitor to review what variables have changed.

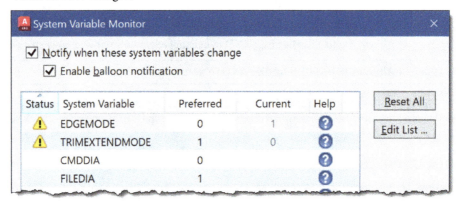

Figure 13.17: The System Variable Monitor changes

6. Use the **UNDO** (*Ctrl + Z*) command to restore the previous graphics.

7. Select the **Trim** command and select both the vertical and horizontal LINE objects as cutting edges. Then, use the *Enter* key to complete the selection.

8. Select the horizontal LINE objects to trim and notice the difference. The vertical LINES are now recognized using their "imaginary" extensions.

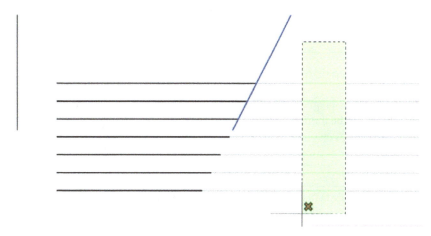

Figure 13.18: TRIM using EDGEMODE results

Note

You will undoubtedly find that the "balloon" notification is a little annoying; so, just run the SYSVARMONITOR command again and turn OFF the **Enable Balloon Notification** while leaving the **Notify when these system variables change** setting turned on.

Turn this setting back on when things are not working normally in AutoCAD to discover what has changed.

In the next section, we will concentrate on the system variables and commands that control linetype scales.

Using linetype scales

Scaling your linetypes in project files should be fairly automatic these days, but I find that many users don't know all the features that control the linetype scaling. So, let's investigate how this all works.

LTSCALE

First, let's review the LTSCALE command, which most of you already know about. But just in case, let's review it.

The LTSCALE command

The LTSCALE command defines the global scale applied to all linetypes in the drawing file. By default, the LTSCALE is set to a scale factor of 1 using the out-of-the-box template files. Many users prefer to set this to .5 for more visible linetype repetitions. The smaller this number, the more repetitions are displayed.

LTSCALE	Command Locations
Command Line	LTSCALE (LTS)

The LTSCALE system variable

The LTSCALE system variable provides a method for setting the preferred linetype scale in your drawing. This variable can be monitored using the System Variable Manager to notify you when the file LTSCALE command has changed the scale factor.

LTSCALE
This defines the global scale factor for linetypes in the drawing file.
Type: Real
Saved in: Drawing
Initial Value: 1.0000

PSLTSCALE

The PSLTSCALE system variable controls the linetype scaling of objects on a per-viewport basis. When a viewport scale is defined, the PSLTSCALE adjusts the display of the linetypes in the viewport based on that scale.

When you change the scale of a viewport, the scale change of linetype objects is not automatically displayed. You must use the REGEN or REGENALL commands to update the display in the viewport.

PSLTSCALE	
This controls the scale of linetypes in the layout tabs, where the scale is adjusted by each viewport scale.	
Type: String	
Saved in: Drawing	
0	OFF. The linetype scale is not adjusted based on the viewport scale. The linetype scale is determined by the model space LTSCALE factor.
1 (default)	ON. The linetype scale is adjusted based on the viewport scale.

MSLTSCALE

The MSLTSCALE system variable controls the linetype scaling of objects on the model space tab.

MSLTSCALE	
This controls the scale of linetypes in the model space tab where the scale factor is determined by the current annotation scale.	
Type: Integer	
Saved in: Drawing	
0	OFF. Linetype scales are not controlled by the annotation scale.
1 (default)	ON. Linetype scales are controlled by the annotation scale.

CELTSCALE

The CELTSCALE system variable controls the linetype scaling of an individual object in the drawing. This scale factor is applied to the current LTSCALE setting.

CELTSCALE	
This controls the scale of an individual object in the drawing and is a multiplier of LTSCALE.	
Type: Real	
Saved in: Drawing	
Initial Value: 1.0000	
0	OFF. Linetype scales are not controlled by the annotation scale.
1 (default)	ON. Linetype scales are controlled by the annotation scale.

LTGAPSELECTION

Use the LTGAPSELECTION system variable to snap or select a gap in a non-continuous linetype.

LTGAPSELECTION	
This controls your ability to select or snap to gaps in non-continuous linetypes.	
Type: Integer	
Saved in: Register	
Initial Value: 1	
0	OFF. You cannot snap or select on a gap in the linetype.
1 (default)	ON. You can snap or select a gap in the linetype.
Note: Hardware acceleration and High-Quality Geometry must be turned ON for the LTGAPSELECTION to function.	

Now, let's look at how these commands and system variables work in the real drawing file:

1. Continue using the 13-3_AutoCAD Configuration.dwg file.

2. Using the In-Canvas View Controls, restore **Custom Model Views | 3-Linetype Scales** named view.

3. Using the status bar, verify that the current ANNOTATION SCALE is set to **1:1**.

Figure 13.19: Initial annotation scale

4. Change the current annotation scale to **1:2**, and you will see the TEXT objects update, but not the linetypes.

5. Using the Command Line, key in the REGEN command to update the display of the linetypes for 1:2. The linetypes are scaled up for 1:2.

6. Change the current annotation scale back to **1:1**.

7. Using the Command Line, key in the LTSCALE command and change it to .5. The linetype scales are adjusted, but not the text or any other annotation objects. Change the **LTSCALE** back to 1.

8. Select the **5-Linetype Scales** layout tab and *double-left-click* in the **SCALE 1:1** viewport to activate model space. You might need to run the REGEN command to update the linetype scales.

9. Using the status bar, verify that the current annotation scale for this viewport is set to **1:1**.

10. *Left-click* in the **SCALE 1:2** viewport and verify that the current annotation scale for this viewport is set to **1:2**.

11. Change the current annotation scale to **1:3**, and you will see the TEXT objects update, but not the linetypes.

12. Using the Command Line, key in the REGEN command to update the display of the linetypes for 1:3. The linetypes are scaled up for 1:3.

13. *Double-left-click* in paper space or click the CHANGE SPACE button PAPER in the status bar to switch to paper space. This button is particularly useful if you are zoomed in on the sheet layout and cannot see any of the "gray" paper space areas.

14. *Key in the* REGEN *command to update the text label* **SCALE 1:2** *to read the new scale of* **1:3**.

Now that you have briefly experimented with the linetype scale commands, you should notice that regardless of the viewport scale, the linetypes look identical "automatically" after the REGEN command.

When these variables are set how you prefer, you rarely have to worry about linetype scales again. But when things go wrong, you must know all of these settings.

Advanced viewport options

In this section, we will examine how to use some of the more obscure features of using viewports in AutoCAD.

Merge layout viewports

Often, during a project, our details get scattered across multiple layouts because when they are created at an earlier stage, we don't always know what details we will ultimately end up with. Many times, one detail is deleted altogether, and new details are added.

In this exercise, we will learn to merge details from one layout to another.

1. Open the 13-4_VIEWPORT Commands.dwg file.

2. Select the **Details Sheet 1** layout tab, and using the **Express Tools** ribbon and the **Layout** panel, select the **Merge Layout** command. 🔳

3. Using the **Layout Merge** dialog, select the **Details Sheet 1** and **Details Sheet 2** layouts and click **OK** to close the dialog.

4. Select the **Details Sheet 1** layout when asked to specify the destination Layout.

5. Answer YES to delete the unused layout to remove any duplication.

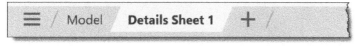

Figure 13.20: Merge layouts results

The view expands to display both layouts in paper space.

6. Select the two merged viewports and their TITLE TEXT objects and MOVE them to the **Details Sheet 1** layout.

7. Delete any unwanted graphics from the merged **Details Sheet 2** graphics.

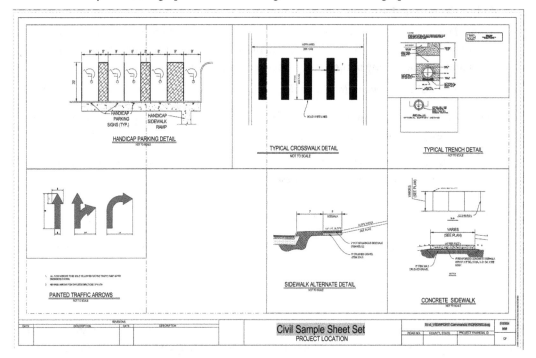

Figure 13.21: Merge layouts results

In the next exercise, we will learn how to combine two viewports into a single viewport using the VIEWPORT CLIP (VPCLIP) command.

VPCLIP	Command Locations
Ribbon	Layout | Layout Viewports | Clip
Right-click Menu	Select a Viewport | Clip
Command Line	VPCLIP (VPC)

Viewport Clip

Often, we need to reshape a viewport to accommodate additional graphics that have been added during the project. To accomplish this, use the VPCLIP command to redefine the shape of an existing viewport:

1. Continue using the 13-4_VIEWPORT Commands.dwg file.

2. Select the **Details Sheet 1** layout and zoom in on **3-TYPICAL TRENCH DETAIL**.

This detail has two viewports using two different scales, and we want to combine them into a single viewport.

3. Using the **Layout** ribbon and the **Layout Viewports** panel, select the **Clip** command.

4. Select the larger viewport POLYLINE and use the *Enter* key to complete the selection.

> **Note**
>
> If you are using object cycling, you need to select the POLYLINE object, not the VIEWPORT.

5. Trace around the two viewports as needed and close the viewport when completed. The resulting viewport will be displayed using the scale from the initial viewport.

6. Use the **Delete** command to remove the smaller viewport now inside the new polygonal viewport.

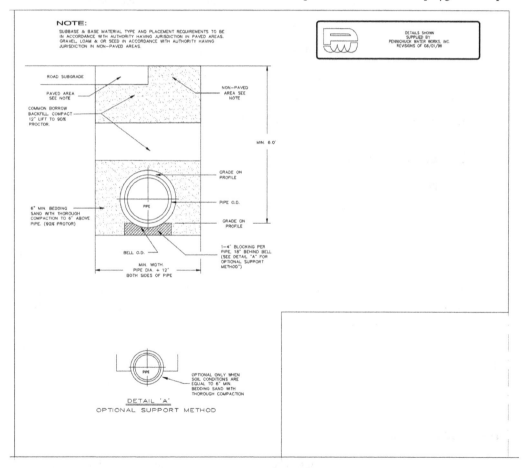

Figure 13.22: VPCLIP results

Change space

You can use the CHANGE SPACE command to transfer objects from model space to paper space or vice versa.

1. Continue using the `13-4_VIEWPORT Commands.dwg` file.
2. Select the **Detail Sheet 1** layout tab and zoom in on **2-TYPICAL CROSSWALK DETAIL**.

Here, you see the TITLE objects are in model space and not in paper space where we want them. We need to transfer these TITLE objects to paper space while leaving the detail graphics in model space. Using CUT and COPY with the clipboard will not account for the scaling differences:

1. *Double-left-click* in the viewport to activate model space.
2. Using the **Home** ribbon and the **Modify** panel, select the **Change Space** command.
3. Select the **TITLE** objects and press *Enter* to complete the command.

The title is now in paper space where it should be. However, its size is not the same as our other titles.

1. Using the QAT toolbar, select the **Match Properties** command.
2. Select the TITLE object, **1-HANDICAP PARKING DETAIL**, first, since this text object is the correct size, and select **2-TYPICAL CROSSWALK DETAIL** second to fix the properties.

TRIM or EXTEND to TEXT objects

After moving the TITLE objects to paper space, the LINE objects for the CROSSWALK title text are too long. We will use the TRIM command and trim the TEXT object. In AutoCAD 2021 and forward, you must use the STANDARD trim mode to work like the previous versions of the TRIM command.

1. Select the **Trim** command, select the **Mode** command option, and then select the **Standard** command option.
2. Select the TEXT object above the lines to define it as the "cutting" object and press *Enter*. Next, drag a **Crossing Window** over both the LEFT endpoint of the horizontal line and the RIGHT endpoint of the horizontal line to trim the LINE to the TEXT object.

> **Note**
> TEXT objects should be MTEXT objects to work with the TRIM and EXTEND commands.

Rotated viewports

We often need to rotate the contents of a viewport in paper space to align the graphics and improve their fit on the paper view. Several methods exist, some more efficient than others.

Yes, you can rotate the viewport and then modify its shape, but that takes a lot of steps. I prefer to use one of the following methods.

MVIEW command

Use the MVIEW command to quickly create a viewport and fit the drawing graphics in the new viewport. This is much easier than using the VPORT command dialog:

1. Continue using the `13-4_VIEWPORT Commands.dwg` file.

2. Select the **Detail Sheet 1** layout tab and zoom into the **4-PAINTED TRAFFIC ARROWS** portion of the layout.

3. Using the Command Line, key in the `MVIEW` command and draw a new viewport next to the PAINTED TRAFFIC ARROWS viewport.

4. *Double-left-click* in the new viewport and zoom in on the TRAFFIC ARROWS graphics.

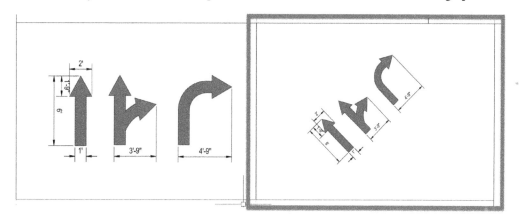

Figure 13.23: MVIEW results

Next, we want to use the VIEWTWIST command to determine the rotation angle of an existing viewport.

Viewport angle

Use the VIEWTWIST command to find the angle of a rotated viewport.

VIEWTWIST

The VIEWTWIST command provides a quick and easy method for determining the rotation angle of an existing viewport:

1. Continue using the `13-4_VIEWPORT Commands.dwg` file.

2. Select the **Details Sheet 1** layout tab and zoom in on the **4-PAINTED TRAFFIC ARROWS** detail.

3. *Double-left-click* in the viewport to activate model space. Notice that the UCSICON displays the view as rotated, but it does not display the specific rotation angle.

4. While still in the viewport, key in the VIEWTWIST command; the rotation angle is displayed in the command line.

```
Command: *Cancel*
Command: VIEWTWIST
VIEWTWIST = 315 (read only)
```

Figure 13.24: VIEWTWIST results

Now that you know the rotation angle, you can rotate the viewport and realign its shape. However, I think that is too difficult and involves too many steps. Try using the DVIEW command to "twist" the contents of the viewport.

The DVIEW command

The DVIEW command provides several methods for changing the parallel or perspective view in a viewport. Many of the command options are primarily used in perspective rotations, but the TWIST command option can be used to twist or tilt a view about a direct line of sight:

1. Continue using the 13-4_VIEWPORT Commands.dwg file.

2. *Double-left-click* in the rotated TRAFFIC ARROWS viewport to activate model space and key in the View Cube.

3. Press the *Enter* key to accept the **use DVIEWblock** option and select the **Twist** command option.

4. Key in the angle -45 and press the *Enter* key to complete the rotation.

5. Use the **UNDO** (*Ctrl* + *Z*) command to remove this rotation method.

We can eliminate the need to know what angle to "twist" the graphics in the viewport by using ISOLATE OBJECTS and the UCSICON commands.

The UCSICON command

My favorite method for rotating the contents of a viewport is to use the UCSICON along with the MOVE AND ALIGN command option.

First, we need to isolate the objects we want in the view to prevent the ZOOM EXTENTS from including other objects:

1. Continue using the 13-4_VIEWPORT Commands.dwg file.

2. *Double-left-click* in the rotated TRAFFIC ARROWS viewport to activate model space and select the graphics you want to include in the viewport. For this exercise, we will select the three traffic arrow graphics and their dimensions.

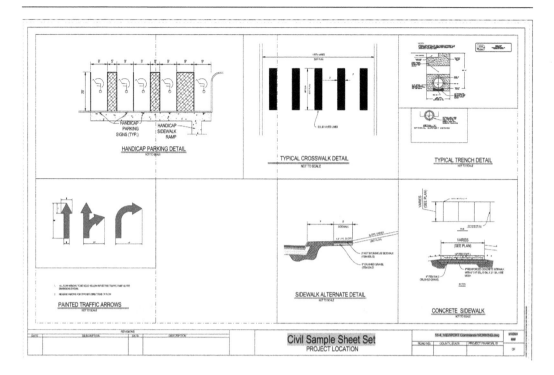

Figure 13.25: Traffic arrows detail

3. *Using the status bar, left-click on the* ISOLATE OBJECTS *button* ⌷○ *and select the* **Isolate Objects** *command. You can also use a right-click to access the* **Isolate** *command.*

    This will turn off all but the objects we want to concentrate on. You should see the remaining model space objects disappear in the other viewports.

    It is important to isolate the objects for the next few steps because isolating these objects improves the view performance when UCSFOLLOW automatically performs a ZOOM EXTENTS.

4. *Using the status bar, unlock the current viewport using the VIEWPORTS LOCK icon. Remember, BLUE is on* 🔒 *and GRAY is off* 🔓.

5. *Left-click* on the UCSICON and *hover* on the origin grip to access the **Move and Align** command.

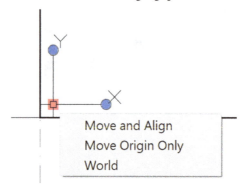

Figure 13.26: UCSICON Move and Align

6. Move the UCSICON to **P1** for the ORIGIN, select the **X-AXIS** GRIP, and align it to **P2**.

7. The view in the viewport automatically rotates to align with the X-AXIS of the UCSICON with the X-AXIS of the viewport. Use the *Esc* key to de-select the UCSICON.

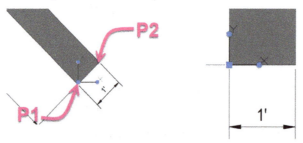

Figure 13.27: UCSICON Move and Align graphics and results

8. *Using the status bar, left-click on the* UNISOLATE OBJECTS *button* and *select the* **End Object Isolation** *command. You can also use a right-click to access the* **Isolate** *command.*

9. While still in the viewport, verify that the VIEWPORT SCALE is still set to **3/8" = 1'-0"** and use the **PAN** command to align the objects in the viewport.

10. *Using the status bar, lock the current viewport using the VIEWPORTS LOCK icon.*

11. *Double-left-click* in paper space to move to paper space.

Now, you know a couple of different methods for rotating a viewport and aligning your project graphics to paper space.

UCSFOLLOW

Use the UCSFOLLOW system variable to control how your view reacts when changing from one UCS to another.

UCSFOLLOW	
This creates a PLAN view when you change from one UCS to another. The variable is saved separately for each viewport.	
Type: Integer	
Saved in: Drawing	
0 (default)	The current view is not affected by changing the UCS.
1	The current view is affected by changing the UCS.
Note:	The UCSFOLLOW setting is ignored when using paper space.

In the next exercise, we will gain a better understanding of how the UCSFOLLOW system variable works. Let's try this system variable out:

1. Open the 13-5_BONUS Commands.dwg file.

2. Using the In-Canvas View Controls, restore **Custom Model Views | 1-UCSFOLLOW** named view.

3. Using the Command Line, key in the UCSFOLLOW system variable and change the setting to 1.

4. Using the Command Line, key in the UCSICON system variable and select the **Origin** command option to display the UCS ICON at 0,0,0.

5. Using the View Cube, use the **WCS** *drop-down list* to access the available named UCS and select the **Top Slanted Face** UCS.

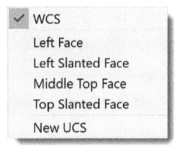

Figure 13.28: Named UCS

The view rotates to follow the selected UCS and performs a ZOOM EXTENTS.

6. Select the **Middle Flat Face** UCS, and again, the view rotates to align automatically.

If you prefer to work in an isometric view and rotate only the UCS alignment, set the UCSFOLLOW system variable back to 0 to disable this feature.

7. Using the View Cube, use the **WCS** *drop-down list* to access the available named UCS and select the **Top Slanted Face** UCS.

This time, the view does not rotate to follow with the selected UCS.

In the next section, we will look at the commands that are not being used enough by newer users.

More bonus commands

In this section, we will examine a few more commands that users often overlook. Some of these commands are only available using the command line, and others are only on the *right-click* menus.

The first command to take a look at is the DIVIDE command, which is used to place points along an existing object.

DIVIDE

The DIVIDE command places evenly spaced POINTS or BLOCKS along an object. These points can then be used as snappable points to add additional graphics. When POINTS are placed, the PTYPE command controls what style and size of POINT is displayed.

DIVIDE	Command Locations
Ribbon	Home | Extended Draw | Divide
Command Line	DIVIDE (DIV)

PTYPE	Command Locations
Ribbon	Home | Utilities | Point Style
Command Line	PTYPE (PT)

Dividing using points

In this exercise, we will use the dividing into equal distances" DIVIDE command to easily offset lines to divide the space into equal distances:

1. Continue using the 13-5_BONUS Commands.dwg file.
2. Using the In-Canvas View Controls, restore **Custom Model Views \| 2-DIVIDE** named view.
3. Using the **Home** ribbon and the expanded **Utilities** panel, select the **Point Style** command.
4. Using the **Point Style** dialog, select any style other than the first two, which are a SMALL DOT or NO DOT. Both of these styles are very difficult to see on screen.

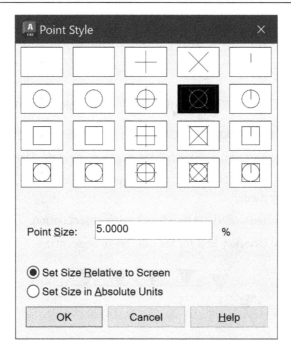

Figure 13.29: Point styles

> **Note**
>
> Use the first two point styles to maintain points that will remain in the drawing but will be invisible on the screen or when plotted.

The defaultdividing into equal distances" point style size is five percent of the screen size, however, you can change it to represent a specific unit size in absolute units:

1. Click **OK** to Close the **Point Style** dialog.

2. Using the **Home** ribbon and the expanded **Draw** panel, select the **Divide** command.

3. Select the TOP POLYLINE object and key in 5 for the number of segments.

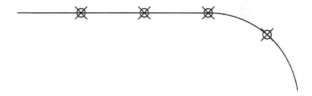

Figure 13.30: Divide polyline

4.   Use the dividing into equal distances" NODE Osnap to snap to the new POINT objects on the POLYLINE.

Dividing using blocks

In this exercise, we will use a BLOCK object instead of a POINT object with the DIVIDE command:

1.   Continue using the 13-5_BONUS Commands.dwg file.

2.   Using the **Home** ribbon and the expanded **Draw** panel, select the **Divide** command.

3.   Select the MIDDLE POLYLINE object, then select the **Block** command option, and key in ARROW for the block name.

4.   Key in N when presented with the **Align block with object** option.

5.   Using the Command Line, key in 7 for the number of segments.

Figure 13.31: Divide with Block not Aligned

6.   Using the **Home** ribbon and the expanded **Draw** panel, select the **Divide** command.

7.   Select the BOTTOM POLYLINE object, then select the **Block** command option, and key in ARROW for the block name.

> **Note**
> You can also use the up arrow key on the keyboard to recall the previous key.

8.   Key in Y when presented with the **Align block with object** option.

9.   Using the Command Line, key in 7 for the number of segments.

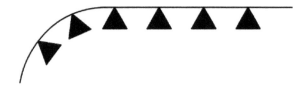

Figure 13.32: Divide with Block Aligned

In the next exercise, we will learn how to blend two or more objects using the BLEND CURVE command.

BLEND CURVE

Similar to the FILLET and CHAMFER commands we all use every day, the BLEND command provides a method for transitioning between 2D objects. The BLEND command creates SPLINE objects to join the 2D objects.

BLEND	Command Locations
Ribbon	Home	Modify	Fillet	Blend Curve
Command Line	BLEND (BLE)

In this exercise, we will learn to use the BLEND command to automatically connect objects with smooth curves:

1. Continue using the 13-5_BONUS Commands.dwg file.

2. Using the In-Canvas View Controls, restore **Custom Model Views | 3-BLEND** named view.

3. Using the **Home** ribbon and the **Modify** panel, select the FILLET *drop-down list* and select the **Blend Curve** command.

4. Select the FIRST ARC object near **P1** and select the SECOND ARC object near **P2**.

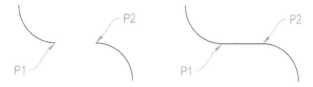

Figure 13.33: BLEND CURVE results 1

Next, we will blend two objects that are not aligned:

1. Select the **Blend Curve** command again, or use the spacebar to recall the previous command.

2. Select the THIRD ARC object near **P3** and select the FOURTH ARC object near **P4**.

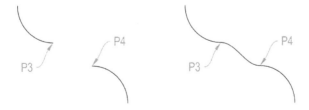

Figure 13.34: BLEND CURVE results 2

3. Select the **Blend Curve** command again, then select the TOP POLYLINE object and select the MIDDLE POLYLINE to create the following shapes.

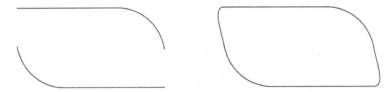

Figure 13.35: BLEND CURVE results 3

Here is another set of shapes created using the BLEND command.

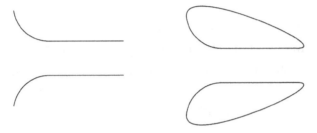

Figure 13.36: BLEND CURVE results 4

Summary

In this chapter, we learned how to improve our drawing performance using the PERFORMANCE ANALYZER, OVERKILL, and PURGE commands. In addition to those utilities, we learned how to improve our production performance by controlling our linetype scales, system variables, and viewports.

This concludes the last chapter of this book. I hope you enjoyed the journey through these tips and techniques of AutoCAD and you learned about more than a few new features that you can successfully apply in your daily AutoCAD adventure.

Index

www.packtpub.com

Subscribe to our online digital library for full access to over 7,000 books and videos, as well as industry leading tools to help you plan your personal development and advance your career. For more information, please visit our website.

Why subscribe?

- Spend less time learning and more time coding with practical eBooks and Videos from over 4,000 industry professionals

- Improve your learning with Skill Plans built especially for you

- Get a free eBook or video every month

- Fully searchable for easy access to vital information

- Copy and paste, print, and bookmark content

Did you know that Packt offers eBook versions of every book published, with PDF and ePub files available? You can upgrade to the eBook version at packtpub.com and as a print book customer, you are entitled to a discount on the eBook copy. Get in touch with us at customercare@packtpub.com for more details.

At www.packtpub.com, you can also read a collection of free technical articles, sign up for a range of free newsletters, and receive exclusive discounts and offers on Packt books and eBooks.

Other Books You May Enjoy

If you enjoyed this book, you may be interested in these other books by Packt:

Practical Autodesk AutoCAD 2021 and AutoCAD LT 2021

Yasser Shoukry, Jaiprakash Pandey

ISBN: 978-1-78980-915-2

- Understand CAD fundamentals using AutoCAD's basic functions, navigation, and components
- Create complex 3d solid objects starting from the primitive shapes using the solid editing tools
- Working with reusable objects like Blocks and collaborating using xRef
- Explore some advanced features like external references and dynamic block
- Get to grips with surface and mesh modeling tools such as Fillet, Trim, and Extend
- Use the paper space layout in AutoCAD for creating professional plots for 2D and 3D models
- Convert your 2D drawings into 3D models

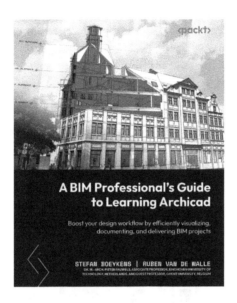

A BIM Professional's Guide to Learning Archicad

Stefan Boeykens, Ruben Van de Walle

ISBN: 978-1-80324-657-4

- Create an architectural model from scratch using Archicad as BIM software
- Leverage a wide variety of tools and views to fully develop a project
- Achieve efficient project organization and modeling for professional results with increased productivity
- Fully document a project, including various 2D and 3D documents and construction details
- Professionalize your BIM workflow with advanced insight and the use of expert tips and tricks
- Unlock the geometric and non-geometric information in your models by adding properties and creating schedules to prepare for a bill of quantities

Packt is searching for authors like you

If you're interested in becoming an author for Packt, please visit `authors.packtpub.com` and apply today. We have worked with thousands of developers and tech professionals, just like you, to help them share their insight with the global tech community. You can make a general application, apply for a specific hot topic that we are recruiting an author for, or submit your own idea.

Share Your Thoughts

Now you've finished *AutoCAD 2025 Best Practices, Tips, and Techniques*, we'd love to hear your thoughts! Scan the QR code below to go straight to the Amazon review page for this book and share your feedback or leave a review on the site that you purchased it from.

`https://packt.link/r/1837636729`

Your review is important to us and the tech community and will help us make sure we're delivering excellent quality content.

Download a free PDF copy of this book

Thanks for purchasing this book!

Do you like to read on the go but are unable to carry your print books everywhere?

Is your eBook purchase not compatible with the device of your choice?

Don't worry, now with every Packt book you get a DRM-free PDF version of that book at no cost.

Read anywhere, any place, on any device. Search, copy, and paste code from your favorite technical books directly into your application.

The perks don't stop there, you can get exclusive access to discounts, newsletters, and great free content in your inbox daily

Follow these simple steps to get the benefits:

1. Scan the QR code or visit the link below

https://packt.link/free-ebook/9781837636723

2. Submit your proof of purchase
3. That's it! We'll send your free PDF and other benefits to your email directly

www.ingramcontent.com/pod-product-compliance
Lightning Source LLC
Chambersburg PA
CBHW060647060326
40690CB00020B/4544